Communications
in Computer and Information Science 1304

More information about this series at http://www.springer.com/series/7899

Vladimir Jordan · Nikolay Filimonov ·
Ilya Tarasov · Vladimir Faerman (Eds.)

High-Performance Computing Systems and Technologies in Scientific Research, Automation of Control and Production

10th International Conference, HPCST 2020
Barnaul, Russia, May 15–16, 2020
Revised Selected Papers

 Springer

Editors
Vladimir Jordan (iD)
Altai State University
Barnaul, Russia

Ilya Tarasov (iD)
MIREA – Russian Technological University
Moscow, Russia

Nikolay Filimonov (iD)
Trapeznikov Institute of Control
Problems of RAS
Moscow, Russia

Vladimir Faerman (iD)
Tomsk State University of Control
Systems and Radioelectronics
Tomsk, Russia

ISSN 1865-0929   ISSN 1865-0937   (electronic)
Communications in Computer and Information Science
ISBN 978-3-030-66894-5   ISBN 978-3-030-66895-2   (eBook)
https://doi.org/10.1007/978-3-030-66895-2

This Springer imprint is published by the registered company Springer Nature Switzerland AG
The registered company address is: Gewerbestrasse 11, 6330 Cham, Switzerland

Preface

The 10th International Conference on High-Performance Computing Systems and Technologies in Scientific Research, Automation of Control and Production (HPCST 2020) took place at Altai State University on May 15–16, 2020. Altai State University (AltSU) is located in the center of Barnaul city - the capital of the Altai region in the southwestern part of Siberia.

HPCST is a regular scientific meeting that has been held annually since 2011. It attracts specialists in the various fields of modern Computer and Information Science, as well as their applications in the automation of control and production, in mathematical modeling and computer simulation of processes and phenomena in the natural sciences by means of parallel computing.

The goal of the conference is to present state-of-art approaches and methods for solving contemporary scientific problems and to exchange the latest research results obtained by scientists from both universities and research institutions. All the reported results are valuable contributions to the field of applied information and computer science.

Sessions of the conference are devoted to the relevant scientific topics:

- architecture and design features of high-performance computing systems;
- digital signal processors (DSP) and their applications;
- IP cores for field-programmable gate arrays (FPGA);
- technologies for distributed computing using multiprocessors;
- GRID technologies and cloud computing and services;
- high-performance and multi-scale predictive computer simulation;
- control automation and mechatronics.

Since the conference was introduced at the International level back in 2017, more than 100 researchers from Russia, China, Ukraine, Kazakhstan, Kyrgyzstan, Uzbekistan and Tajikistan have participated in the conference. The average number of participants in a single year is about 40.

Due to the global pandemic of COVID-19, the conference was postponed from the original date in April to the middle of May. Despite this measure, travel restrictions were still in force in Russia, so the committee made the decision to hold the meetings in a semi-virtual format. The local audience from Barnaul, its suburbs and surrounding cities participated *in presentia*. At the same time, the distant audience participated remotely via Zoom. Specifically for the purpose, several rooms in the AltSU main building were supplied with equipment for videoconferencing.

The conference was attended virtually or *in presentia* by 112 scholars with more than forty accepted presentations. Only the most significant reported studies were thoroughly reviewed and included in these post-proceedings. The vast majority of the conference papers that were not selected by the committee have been published as regular proceedings.

Fourteen full papers featuring original studies in the field of computing, mathematical simulation and control science comprise this volume. The papers cover such topics as:

- Hardware for High-Performance Computing and Its Applications;
- Information Technologies and Computer Simulation of Physical Phenomena;
- Computing Technologies in Discrete Mathematics and Decision Making;
- Information and Computing Technologies in Automation and Control Science.

To select the best papers among those presented at the conference the following procedure was applied.

1. Section chairs made a short list of the most significant original reports which had clear potential to be extended to a full-paper format.
2. The editorial board, comprising the session chairs and corresponding editor, made a list of 15 items. Authors of those manuscripts were contacted and asked to extend their papers and resubmit them for review.

To ensure that after the review the volume would meet both qualitative and quantitative requirements, several invitations to submit a full paper for review were distributed among the committee, reviewers and collaborators.

Every paper without any exceptions was reviewed by at least three experts. A single-blind review method was applied. The review criteria are revealed below:

1. Technical content;
2. Originality;
3. Clarity;
4. Significance;
5. Presentation style;
6. Ethics.

Among the 14 papers in the volume, 11 papers were selected among the 46 accepted reports. The other 3 accepted papers were selected from the 11 manuscripts suggested by the committee and reviewers. Therefore, the acceptance rate was a little less than 25%.

The organizing committee would like to express our sincere appreciation for the organizational support of the administration of Altai State University and to the staff of the Institute of Digital Technology, Electronics and Physics of Altai State University. Only the outstanding effort of the technical staff made the conference possible at a time of travel restrictions.

The editors would like to express their deep gratitude to the Springer editorial team for the provided opportunity to publish the best papers as post-proceedings and for their great work on this volume.

November 2020   Vladimir Jordan
                                                               Nikolay Filimonov
                                                                    Ilya Tarasov
                                                               Vladimir Faerman

Organization

General Chair

Vladimir Jordan   Altai State University, Russia

Program Committee Chairs

Ella Shurina   Novosibirsk State Technical University, Russia
Ilya Tarasov   Russian Technological University, Russia
Nikolay Filimonov   V.A. Trapeznikov Institute of Control Sciences RAS, Russia

Organizing Committee

Vasiliy Belozerskih   Altai State University, Russia
Vladimir Faerman   Tomsk State University of Control Systems
           and Radioelectronics, Russia
Alexander Kalachev   Altai State University, Russia
Vladimir Pashnev   Altai State University, Russia
Viktor Sedalischev   Altai State University, Russia
Yana Sergeeva   Altai State University, Russia
Igor Shmakov   Altai State University, Russia
Petr Ulanov   Altai State University, Russia

Program Committee

Viktor Abanin   Biysk Technological Institute, Russia
Darya Alontseva   D. Serikbayev East Kazakhstan Technical University, Kazakhstan
Valeriy Avramchuk   Tomsk State University of Control Systems
           and Radioelectronics, Russia
Sergey Beznosyuk   Altai State University, Russia
Alexander Filimonov   MIREA - Russian Technological University, Russia
Pavel Gulyaev   Yugra State University, Russia
Ishembek Kadyrov   Kyrgyz National Agrarian University, Kyrgyzstan
Alexander Kalachev   Altai State University, Russia
Vladimir Khmelev   Polzunov Altai State Technical University, Russia
Vladimir Kosarev   Khristianovich Institute of Theoretical and Applied
           Mechanics SB RAS, Russia
Lyudmilla Kveglis   Sarsen Amanzholov East Kazakhstan State University, Kazakhstan

Roman Mescheryakov	V.A. Trapeznikov Institute of Control Sciences RAS, Russia
Aleksey Nikitin	Altai State University, Russia
Viktor Polyakov	Altai State University, Russia
Oleg Prikhodko	Al-Farabi Kazakh National University, Kazakhstan
Sergey Pronin	Polzunov Altai State Technical University, Russia
Alisher Saliev	Kyrgyz State Technical University, Kyrgyzstan
Viktor Sedalischev	Altai State University, Russia
Vitaliy Titov	Southwest State University, Russia
Aleksey Yakunin	Polzunov Altai State Technical University, Russia

External Reviewers

Olga Berestneva	Tomsk Polytechnic University, Russia
Fedor Garaschenko	Kyiv National University n.a. Shevchenko, Ukraine
Anatoliy Gulay	Belarusian National Technical University, Belarus
Gambar Guluev	Institute of Control Systems of the National Academy of Sciences of Azerbaijan, Azerbaijan
Semyon Kantor	Polzunov Altai State Technical University, Russia
Bibigul Koshoeva	Kyrgyz State Technical University, Kyrgyzstan
Eugeniy Kostuchenko	Tomsk State University of Control Systems and Radioelectronics, Russia
Andrey Kutyshkin	Yugra State University, Russia
Elena Luneva	Tomsk Polytechnic University, Russia
Andrey Malchukov	Tomsk State University of Control Systems and Radioelectronics, Russia
Leonid Mikhaylov	Al-Farabi Kazakh National University, Kazakhstan
Nomaz Mirzaev	University of Information Technologies n.a. al-Khorezmi, Uzbekistan
Evgeniy Mytsko	Tomsk Polytechnic University, Russia
Stepan Nebaba	Tomsk Polytechnic University, Russia
Pavel Nedyak	Tomsk State University of Control Systems and Radioelectronics, Russia
Natalya Nikonova	Tomsk State University, Russia
Maksim Pushkaryov	Tomsk Polytechnic University, Russia
Andrey Russkov	Yandex, Russia
Alexey Saveliev	Tomsk Polytechnic University, Russia
Kseniya Zavyalova	Tomsk State University, Russia

Contents

Hardware for High-Performance Computing and Its Applications

FPGA-Based SOC Architecture for Fog and Edge Computing Applications

Ilya Tarasov$^{(\boxtimes)}$ ⓘ and Dmitry Potekhin

MIREA – Russian Technological University, Vernadsky Avenue 78, 119454 Moscow, Russia
tarasov_i@mirea.ru

Abstract. The article examines an example of the architecture of a digital device of the 'system on a chip' class, which implements data collection from sensors of physical quantities and their preliminary processing using integral transformations. This approach is in line with the current trends in fog and edge computing, which provide for distributed data collection, preprocessing and transmission of only conversion results over wireless networks, which can significantly reduce the required traffic. The architecture is designed to accommodate conflicting requirements for high computing performance, the presence of a large number of external interfaces and lower power consumption; therefore, its implementation requires specialization of components in relation to the problem being solved. The architecture variant was implemented on the basis of FPGA with FPGA architecture. Examples of IoT devices include a high-resolution wearable cardiograph with multi-channel myograph capability with WiFi and BlueTooth interfaces, and an acoustic emission meter in power equipment. Devices are based on FPGA Xilinx Spartan-7 with external controllers for wireless interfaces.

Keywords: Internet of Things · Fog computing · SoC · FPGA

1 Introduction

Fog computing and edge computing systems are becoming more widespread as new demands in the field of measurement automation emerge, including both industrial automation and wearable Internet of Things [1–5]. New areas of application of computing and communication devices form new requirements for their characteristics. In particular, the performance and power requirements required for wearable devices may not always be met by widely used general purpose hardware platforms. If we consider the specifics of fog and edge computing, we can draw attention to the fact that we are talking, on the one hand, about high-performance devices that process signals at the point of their receipt, and on the other hand, their mobile nature determines the requirements for low power consumption. Therefore, at the present time, the urgency of creating a specialized element base that has sufficient performance to perform pre-processing of signals, and at the same time has a reduced power consumption, is increasing. The combination of these properties is achievable when developing specialized solutions.

© Springer Nature Switzerland AG 2020
V. Jordan et al. (Eds.): HPCST 2020, CCIS 1304, pp. 3–13, 2020.
https://doi.org/10.1007/978-3-030-66895-2_1

2 Problem Description

When designing data collection systems, an important problem is the overload of communication channels when trying to collect data directly from sensitive elements. The capabilities of modern ADCs allow generating data streams up to several billions of samples per second (Gsamples/sec), which obviously exceeds the capabilities of modern wireless communication channels. In addition, the features of distributed data collection systems imply the presence of many data sources, which significantly exacerbates the problem, since even with a small data flow from one source, the need to connect multiple sources proportionally increases the amount of information transmitted.

Using a layered architecture that includes data hub middle tiers can alleviate the problem, but this approach should avoid the emergence of thick tree networking architectures. This architecture implies an increase in network traffic when moving from the sensor-hub layer to the next layer, for example, the hub-to-PC layer. The concept of fog and edge computing is a timely solution to this problem, since it allows you to organize preliminary data processing at the point of receipt, which in some cases significantly reduces the traffic required for transmission to a higher level. At the same time, the practical implementation of this approach requires the fulfillment of a number of conditions:

- the problem being solved must objectively allow a decrease in traffic transmitted for analysis; this can be the case when a decision on the measurement results is made on the basis of spectral, statistical or other similar characteristics of the measured signal, implying the use of transformations that reduce the amount of data (for example, transformations based on the calculation of integrals or convolutions have such a property);
- the hardware edge computing platform provides the required level of performance to implement the algorithm at the level of the primary network of sensors or a local hub that receives data from a limited number of sensors.

The use of primary concentrators in distributed measuring systems that collect data from a limited number of sensors allows unloading the higher levels of the hierarchy of such a system. You can pay attention to the fact that in this case the characteristics of the primary concentrator must be studied in detail. On the one hand, its performance and the throughput of external interfaces should be sufficient, but on the other hand, redundancy of characteristics is likely to cause an increase in weight and dimensions, energy consumption and cost, and also complicate operation. Therefore, this article discusses the use of FPGAs for building specialized devices designed to work as data concentrators in distributed measuring systems, which also perform preliminary processing in order to reduce traffic to higher system elements. This FPGA application follows the concept of fog/edge computing.

A known disadvantage of FPGAs with FPGA architecture is the worst performance of clock frequency and specific power consumption compared to CPU/GPU due to the presence of configurable connections and the implementation of logical expressions based on programmable truth tables based on static memory (LUT). At the same time, modern FPGAs have a large number of hardware-implemented components on the chip, the characteristics of which are not degraded relative to solutions comparable in the technological process, since these components are not configurable and do not contain redundant switching components. Traditionally, such devices are static dual-port memory blocks (BRAMs), DSP48 multiply-accumulate components, and, for some FPGA families, high-speed serial transceivers (MGTs). Hardware-based components deliver high absolute performance for FPGAs.

Thus, it is required to consider the architecture of a system that is a concentrator of data from a set of external sources. Such a system, implemented in an FPGA, is designed to connect external data sources, perform preliminary signal processing in order to reduce the amount of transmitted data and connect external interfaces for integrating a data concentrator into a higher-level system. At the same time, it is necessary to pay attention to the efficient use of FPGA hardware components and to compensate for the negative effects from configurable logic cells that have obviously worse technical characteristics. A variant of the considered architecture is given in Fig. 1.

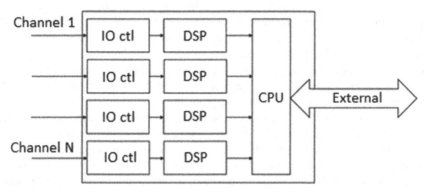

Fig. 1. Hub architecture of the data acquisition subsystem for implementation on the basis of FPGAs.

The ability to configure FPGAs is of interest mainly in the system development process. This is due to the fact that it is advisable to debug and refine digital signal processing algorithms on a workable prototype of the device, while the computing performance, the structure of data processing channels and the interaction of hardware accelerators with the processor can change significantly when obtaining experimental results. Therefore, when developing new systems with requirements that cannot be specified in advance, it is useful to use a reconfigurable hardware platform.

FPGA configuration is complemented by the ability to program an embedded processor, the so-called software processor. While some FPGA families use hardware cores, such as the Cortex-A core in the Xilinx Zynq-7000 and Xilinx Zynq MPSOC families, some of the FPGA's programmable logic cells can be configured to replicate the operation of the processor device.

The characteristics of the soft processor used to control the fog computing device are of significant interest for analysis, mainly at the design stage. Despite the general impression of sufficient computing performance for embedded electronics, power consumption and the ability to integrate components on a chip are significant for the architectures considered. Therefore, the development of specialized digital devices of the 'system on a chip' class is of practical interest from the point of view of eliminating functional redundancy.

Despite the fact that modern soft processors often support the use of cache memory created on the basis of on-chip static FPGA memory, this approach has a number of design and system drawbacks. First of all, the use of external heap memory complicates PCB routing, increases overall power consumption, and generally complicates device design. From the point of view of the system architecture, the appearance of interacting memory components does not allow predicting the exact execution time of critical segments of the code, since a cache miss can occur at an arbitrary point in time. There is a tendency to use Tightly Coupled Memory in real time systems.

For FPGA-based systems with full control of the design at the development stage [1], you can either explicitly indicate the need to use static memory as the main one, or apply an appropriate architectural approach when designing a specialized control processor based on the use of static memory as the only one available to the processor. In combination with a focus on architectures with a high code density (for example, stack architectures with a zero-operand instruction system), this in many cases makes it possible to control hardware accelerators without deploying a standard processor subsystem on a chip, which in this subclass of devices will be functionally redundant.

The control processor of a heterogeneous computing system can have a moderate performance sufficient for solving control problems. This is due to the fact that the control processor is not a critical unit in such a system that determines the performance in the main problem being solved. The hardware costs of implementing the control processor are overhead costs, since the FPGA resources spent on its implementation are diverted from solving basic computational problems. Therefore, the size of the control processor should be reduced while maintaining an acceptable level of functionality and flexibility in implementing the underlying algorithms. The presence of conflicting requirements makes it questionable to use the term 'minimization of hardware costs', since formally a processor architecture can be obtained that meets the mathematical criterion of minimality, but inconvenient for practical work with applied software.

For the control processor, tight integration with hardware accelerators is desirable, including through the creation of specialized system buses. Versatility and scalability are of lower priority because the integration of the processor and accelerators occurs during the system architecture design process. In this case, simplification of communication protocols will have a more noticeable positive effect on the characteristics of the project compared to the potential possibility of changing the system architecture.

3   FPGA Application in Fog and Edge Computing

Modern FPGAs with FPGA architecture provide essential capabilities for the problem being solved. Important are the number of external pins, which determines the throughput of external interfaces and the connectivity of sensors, as well as the total performance of the digital signal processing subsystem based on DSP sections. The comparative characteristics of the Xilinx FPGA families intended for the development of entry-level systems are given below in Table 1 [6]. It can be noted that a number of FPGA-based products provide for the use of external controllers for wireless interfaces, for example, for the Minized board [7].

Table 1. Comparative characteristics of the Xilinx FPGA families intended for the development of entry-level systems.

	Spartan-7	Artix-7	Zynq-7000
Logic cells, k.	6–100	12–215	23–85
Block RAM, Mbit	0.18–4.3	0.72–13.1	1.8–4.9
DSP slices	10–160	40–740	60–220
MGTs	–	2–16	0–16
Max data rate of MGT, Gbit/s	–	6.6	6.6
Programmable inputs/outputs, max.	400	500	328

An important factor affecting the characteristics of such a subsystem is the ability to reduce the amount of data transmitted to a device at a higher level of the hierarchy. Algorithms with this property include integral transforms, for example, the Fourier transform and the wavelet transform. The basis of the wavelet transform is the calculation of the convolution integral of the form:

$$W(\tau, a) = \frac{1}{a} \int_{-t}^{t} x(t - \tau) \cdot \varphi(t)dt.$$

The form of this formula makes it possible to assert that when the integration is replaced by the summation of discrete readings, the amount of information received decreases in proportion to the increase in the number of readings. Therefore, data processing algorithms based on the calculation of the wavelet transform can transmit the values of the convolution integral as the processing results, rather than the samples themselves.

Integral transforms are convenient for FPGA-based implementations, which, as shown in the above Table, even for entry-level families provide the developer with 10 to 740 'multiply and accumulate' hardware modules (DSP slices). With a technically achievable clock frequency of 300–400 MHz (which corresponds to the features of tracing projects with real, rather than peak operating frequencies), the total performance of the digital signal processing subsystem can range from 3 to 300 GMAC/s (billions of operations 'multiply and accumulation' per second).

The calculation of the wavelet spectrum (or 'wavelet density') is performed in the same way as the Fourier spectral density. The difference is that instead of a harmonic series, the wavelet function acts as an analyzing function. In general, only two requirements are imposed on the wavelet function:

1. Localization in the time domain (i.e., the function should decay with distance from the center along the t axis).
2. Absence of a constant component (the integral of the function must be equal to 0).

The use of wavelet analysis opens up wide prospects, since the synthesis of modulating windows for wavelet functions allows one, in a number of cases, to obtain sufficiently high-quality results. By combining the time intervals and the attenuation coefficient of the Gaussian window, it is possible to obtain a set of frequency response of the wavelet function, which differ in the width of the spectrum and the amount of suppression of the Gibbs effect. At the same time, the absence of an analogue of the fast transform for the Morlet wavelet function makes it necessary to calculate the wavelet density by the "direct method" - by repeating the operation "multiply with accumulation" for each of the frequencies of interest. To do this, you can effectively use hundreds and thousands of DSP48 slices, which are available in modern FPGAs.

4   Examples of Systems

Based on the described approach, a number of projects were carried out, implying the connection to the FPGA of a set of sensors that generate a continuous stream of data for analysis. An FPGA-based concentrator was used to perform integral transformations (Fourier analysis, wavelet analysis) and then prepare data for transmission over a wireless interface.

The project of a wearable cardiograph/myograph is a device of the "Internet of Things" class, implemented on the basis of a multichannel ADC, FPGA and WiFi/BlueTooth wireless modules. Medical monitoring devices are currently of interest to a number of researchers [8–10] and are considered as a type of wearable Internet of Things devices. The block diagram of the device is shown in Fig. 2.

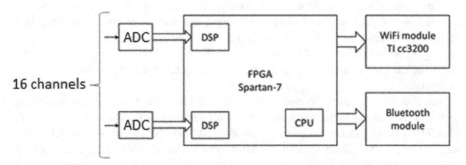

Fig. 2. Block diagram of a cardiograph/myograph with WiFi and BlueTooth wireless interfaces.

Digital signal processing consists in performing band-pass filtering using wavelet filters in the myogram recording mode (measuring muscle activity). The transmission of a raw signal in this mode is of no practical interest, since information on muscle activity is contained in the amplitude of the frequency bands selected for analysis. For signal processing, we used bandpass wavelet filters described by the authors in [11].

In the mode of taking the cardiogram, digital signal processing can be simplified and reduced to filtering high-frequency interference. It can be noted that the so-called. a high-resolution ECG assumes analog-to-digital conversion with a frequency of 2 kHz, so the total load on the wireless interfaces turns out to be small even when transmitting the original data received from the ADC. However, in this case, the FPGA solves another important problem - ensuring the continuity of the measurement process and buffering data. This property is important when using wireless communication channels, in which transmission is carried out in burst mode, and in addition, local fluctuations in the transmission rate are possible due to changes in the reception conditions.

In addition, the independent input of 16 channels of analog-to-digital conversion without using time division multiplexing is a difficult task for a processor with single-threaded instruction execution. The implementation on the basis of FPGA cells and Xilinx XtremeDSP digital signal processing units of a signal preprocessing subsystem with a wavelet transform in real time provided a significant unloading of the control processor, reducing its role in data processing to reading FIFOs with ready conversion results.

The use of an additional digital node that buffers data allows for the collection, processing and transmission without loss of signal fragments in those moments when the connection is unstable or switching to another frequency channel, which may occur for wireless interfaces.

The prototype of the device is shown in Fig. 3. The device uses a Xilinx Spartan-7 FPGA with a logical volume of 25 thousand cells, a single-chip WiFi TI CC3200 controller, and a BlueTooth module. Power is supplied from an external 5 V battery, which allows you to use a wide range of sources designed to power mobile devices. The meter's dimensions allow you to place it in your pocket or on your belt, including for online recording of muscle activity during sports exercises.

The presence of a processor subsystem on the board allows recording signals into flash memory for their subsequent transmission to a higher-level information system for analysis. This makes the device non-critical to the availability of access to a wireless connection, allowing the collection and preliminary analysis of data autonomously. This property is useful for monitoring muscle activity when exercising outdoors, where wireless communication with the nearest WiFi site is likely to be unavailable, and constant use of mobile communication may be undesirable.

Another example of FPGA-based edge computing is the PD meter. A partial discharge is an electrical discharge in insulation, the duration of which is between several tens of nanoseconds. Partial discharge short-term shunts the insulation of high-voltage equipment, which leads to a short-term change in the current in the circuit and is accompanied by acoustic noise. These phenomena can be registered by various methods. Two approaches can be distinguished:

- direct observation of current surges;
- measurements of acoustic noise, for example, using a piezoelectric sensor.

Fig. 3. The cardiograph/myograph with wireless WiFi and BlueTooth interfaces.

Partial discharges appear in the weak point of high-voltage equipment and lead to the gradual development of a defect and destruction of insulation.

Figure 4 shows the spectrum of noise in working high-voltage equipment. Figure 5 shows the noise spectrum in the same equipment in a pre-emergency state. This information can be used to create methods for automated diagnostics, but this requires additional research to obtain an experimental collected base of acoustic noise.

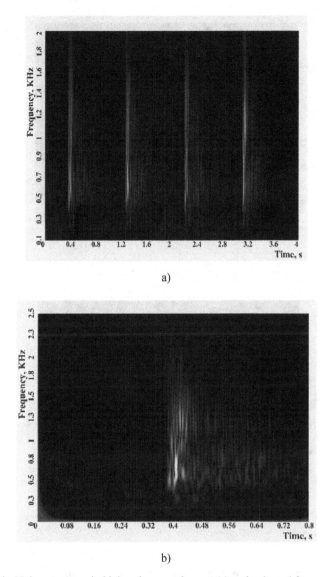

a)

b)

Fig. 4. Noise spectrum in high-voltage equipment (a) and enlarged fragment (b).

The above examples give an idea on the possibilities of foggy and edge calculations in the case of using a specialized element base that performs preliminary signal analysis and transfer only the conversion results. In the examples shown, the stated architectural approach was used to perform multichannel wavelet analysis of signals.

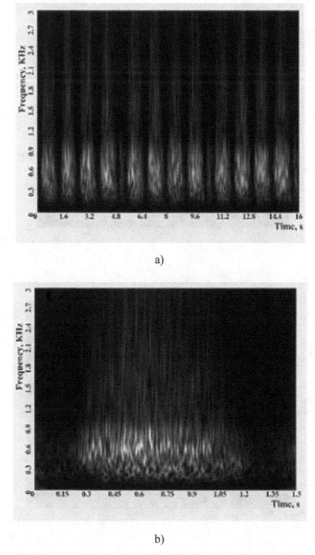

a)

b)

Fig. 5. Noise spectrum in a pre-emergency state in high-voltage equipment (a) and an enlarged fragment (b).

5 Conclusion

The use of FPGA-based signal preprocessing made it possible to use the developed system on a chip to significantly reduce the amount of data required for transmission over wireless communication networks. This approach allows the development of mobile devices for measuring and processing signals in real time with the formation of distributed monitoring networks that can be included in larger information systems - for example, industrial automation systems, medical or sports monitoring systems. The achieved positive effect from the use of FPGAs is a decrease in the amount of data transmitted from the hub, as well as the possibility of clarifying the structure of the device in the process of its design and updating the hardware component during operation.

References

1. Morabito, R., Cozzolino, V., Ding, A.Y., Beijar, N., Ott, J.: Consolidate IoT edge computing with lightweight virtualization. IEEE Netw. **32**(1), 102–111 (2018)
2. Hamm, A., Willner, A., Schieferdecker, I.: Edge computing: a comprehensive survey of current initiatives and a roadmap for a sustainable edge computing development. In: Proceeding of WI2020, pp. 694–709. GITO Verlag. Potsdam, March 2020. https://doi.org/10.30844/wi2 020g1-hamm
3. Pfandzelter, T., Hasenburg, J., Bermbach, D.: From zero to fog: efficient engineering of fog-based IoT applications. Mobile Cloud Computing Research Group Technische Universit at Berlin & Einstein Center Digital Future, 19 August 2020
4. Pfandzelter, T., Bermbach, D.: IoT data processing in the fog: functions, streams, or batch processing? In: Proceedings of the DaMove, pp. 201–206, June 2019
5. Darwish, T.S.J., Bakar, K.A.: Fog based intelligent transportation big data analytics in the Internet of vehicles environment: motivations, architecture, challenges, and critical issues. IEEE Access **6**, 15679–15701 (2018)
6. XILINX Cost-Optimized Portfolio. https://www.xilinx.com/products/silicon-devices/cost-optimized-portfolio.html. Accessed 12 Nov 2020
7. MiniZed Product. http://zedboard.org/product/minized. Accessed 12 Nov 2020
8. Chetelat, O., et al.: Clinical validation of LTMS-S: a wearable system for vital signs monitoring. In: EMBC 2015, pp. 3125–3128 (2015)
9. Segarra, C., Delgado-Gonzalo, R., Lemay, M., Aublin, P.-L., Pietzuch, P., Schiavoni, V.: Using trusted execution environments for secure stream processing of medical data. In: Pereira, J., Ricci, L. (eds.) DAIS 2019. LNCS, vol. 11534, pp. 91–107. Springer, Cham (2019). https://doi.org/10.1007/978-3-030-22496-7_6
10. IHE PCD Technical Committee: medical equipment management (MEM): medical device cyber security. White paper, IHE International, Inc., October 2015
11. Tarasov, I.E., Potekhin, D.S.: Real-time kernel function synthesis for softwaredefined radio and phase-frequency measuring digital systems. Russ. Technol. J. **6**(6), 41–54 (2018). (in Russian). https://doi.org/10.32362/2500-316X-2018-6-6-41-54

Meta-modeling of Space Probe On-Board Computing Complexes

Alexander Lobanov$^{(\boxtimes)}$ ⓘ, Natalia Strogankova ⓘ, and Roman Bolbakov ⓘ

"MIREA" – Russian Technological University, Vernadsky Avenue 78, 119454 Moscow, Russia
aa.lobanoff@ya.ru

Abstract. The study of small bodies of the Solar System (SBSS) is a real and important scientific problem. It demands a space probe landing onto SBSS surface. At the same time, modern space-probe landing procedures make ever-increasing requirements for on-board computing complexes (OBCC). One of the main requirements is an adaptability to the specific mission tasks. It is necessary to reduce the cost of developing the on-board computing complexes with a regard to the growing number of missions to the small bodies in the Solar System. We assume that a meta-modeling is a key to the problem for designing and modeling the on-board computational complexes. The paper describes a meta-modeling and interpretation of a spacecraft on-board hardware and software complexes intended for researching and landing onto small bodies of the solar system. An original approach that could be implemented as a CASE-technology for a full life cycle of OBCC is proposed. The approach is based on a combination a visual algorithmic modeling, a programming language and the SADT methodology. It is designed to meet the requirements of the OBCC ergonomics. The developed meta-model makes it possible to implement modeling of both hardware and software subsystems. Also, it is possible to construe the models automatically into a source code under the meta-data and metalanguage rules. The interpretation is shown in C# programming language. The approach proposed in the paper could significantly optimize the process of spacecraft OBCC design and creation. A hierarchical decomposition of the functional schemes is intended to describe in more detail both the interaction between individual elements and specific submodules of the space probe computational complexes.

Keywords: Meta-modeling · Meta model · Space probe · Semantic element · SADT · On-board computing complex · Space probe landing · Functional model · Visual language

1 The Features of the On-Board Computational Complexes (OBCC) Designing

1.1 The Methodologies Used in the Simulation of OBCC

Nowadays OBCC creation is a complex heuristic procedure especially for specialized space probes (SP). An approach is required that will make it possible to turn a heuristic

© Springer Nature Switzerland AG 2020
V. Jordan et al. (Eds.): HPCST 2020, CCIS 1304, pp. 14–28, 2020.
https://doi.org/10.1007/978-3-030-66895-2_2

procedure into a formal one that requires significantly less resources. The paper provides a feasible solution to the problem. First of all, the authors practically tested the existing methods for describing OBCC. The simulation results allow us to conclude that these methods do not fully meet the specifics of creating such highly specialized computing systems as the space probe's OBCC [1–6]. Some authors use an unsettled term "Cosmionics" by analogy with the aircraft Avionics for that type of systems. However, the approaches successfully employed to create aircraft avionics could not be used for space probes OBCC. The difference is that aircraft avionics is under the constant control of a human operator, when the space probe OBCC operates in an autonomous mode. A real-time monitoring and control of the space probe's on-board computational complexes is not possible.

There is a need to develop a unified approach to design of space probe's on-board computational complexes. It is required to create the OBCC functional modeling methodology. The methodology should be based on a visual modeling and programming language.

In the process of modeling, it became obvious that the SADT methodology is closest to the requirements for creating space probe's complex computing. It was decided to develop additional semantic elements to the IDEF0 notation. Additional semantic elements to the IDEF0 notation for a more accurate description of the design and modeling processes of the OBCC were developed. These semantic elements have been tested on the specific spacecraft subsystems, including:

- an inertial navigation subsystems;
- a computer vision subsystem;
- a laser altimetry subsystem;
- an optical direction finder subsystems;
- a star sensor subsystems.

1.2 General Provisions for OBCC Design

In this article we consider that the on-board computing complex (OBCC) is a set of hardware (HW) and software (SW) that provides an autonomous execution of computing and control processes a space probe in real-time [7].

When developing OBCC, we must provide for its generic applications. Models must imply the principle of reduction descent from high-level abstractions to individual functional elements. It should be noted that the methodology must be hierarchical to provide the model decomposition with any desired degree of process discretization [8–19].

It should be noted that a SP on-board complex functioning is a result of the joint operation of dozens of different software and hardware subsystems. At the design and architecture level the correct operation of the SP onboard complex subsystems must be ensured. Consideration should also be given to the ongoing interaction with external entities of information exchange (EEIE) such as the flight control center.

It should also be noted that in the process of a space mission execution, the data received by the spacecraft is constantly changing. Consequently, an order and a speed of the mission milestones may change. The success of the mission also depends on the speed and correctness of processing the information received [7].

At the stage of OBCC design we favor the IDEF0 methodology as best appropriate for describing the operation of the space probe's OBCC. However, this methodology requires modification to fully meet the OBCC design and functional modeling requirements. While describing the guidance and landing system [7, 19–21], we encountered a number of problems of the use of IDEF0 notation, in particular, and the SADT methodology in general. This approach is incapable to exhaustively describe the features of the operational principle of the OBCC. As a result, it was decided to change the IDEF0 notation to efficiently model the OBCC space probe and provide a program code to describe/use the notation itself. The main feature of the proposed modifications and supplements for the IDEF0 is in their focus on the specific field (space probes' OBCCs). Additionally, the graphic language may be converted into the programming code.

2   The Main Approaches to the Design of the OBCC in the Semantic Environment of Individual Subsystems of the Spacecraft

A space probes' OBCC demands such elements as: logical operators "AND" and "OR", branching after the execution of a process, loops, designation of time frames, as well as service elements necessary to designate relations between processes.

The following requirements must be met:

– a rigor and precision;
– a scalability;
– a compliance with the hierarchical structure;
– a high simulation speed;
– a complexity of a decomposition of functional schemes;
– a completeness of graphical objects for describing processes.

The suggested approach allows to integrate new graphic elements into the original methodology. It is necessary for describing the procedures of on-board computational complexes in a more reliable way. The augmented elements of the methodology turn a projecting process into a trivial procedure.

2.1   Augmented IDEF0 Notation for Modeling an Inertial Navigation Subsystem

The IDEF0 notation is convenient to use when subsystem procedures are sequential and could be presented as a simple algorithm with minimal cycles [16]. When it is not, an additional semantic element is needed. The element is intended to present a transformation of input parameters. It is necessary to model project when they enter the inertial subsystem in the process of modeling has to be used. A developer has to understand a device operating principle. A semantic element "Local computing subsystem" (Fig. 1) provides the ability to describe subsystem processes. It also could be used for complete navigation processes description. The semantic element "Local computing subsystem" (LCS) is integrated into the IDEF0 notation.

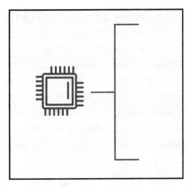

Fig. 1. The suggested semantic element "Local Computing Subsystem" (LCS) of the augmented notation.

The semantic element LCS is intended to describe the processes in the selected subsystem or the onboard equipment with a sufficient accuracy. An example of using the semantic element LCS is presented in Fig. 2. The element more precisely describes a decomposition of the functional block "Calculation of the position of the spacecraft using the inertial subsystem" in relation to the functional block "Determination of the position of the spacecraft in space".

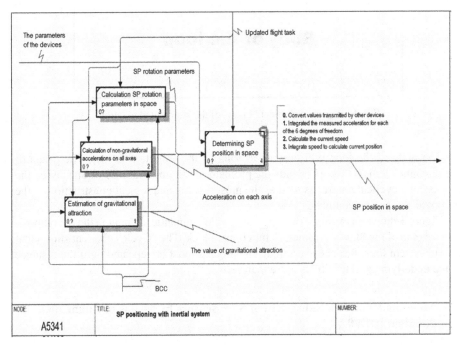

Fig. 2. Decomposition diagram of the functional block "SP positioning with inertial system" A5341 in augmented notation.

The proposed LCS semantic element allows to reduce the technical details of the lower level of a decomposition. It also helps to maintain a logical consistency in the decomposition sequence. The LCS element is useful mechanism for organizing the hierarchy of the functional blocks. The technical details of the operation of the local computing subsystem could be separated. Additionally, the LCS element makes it possible to unambiguously determine the relationship between subsystems of the onboard complex and the logical unit representing it in the IDEF0 notation, since the technical details of the operation of the local computing subsystem can be decomposed into a separate branch of the diagram.

2.2 Meta-modeling of the Laser Altimetry Subsystem Using Augmented Notation

In the process of modeling the laser altimetry subsystem using the IDEF0 notation, the need to supplement the notation with a "Loop" semantic element was also identified [22]. The "Loop" element allows displaying cyclic processes with a specific condition. Figure 3 shows the semantic element "Loop". There is a condition of the loop at the top of the element.

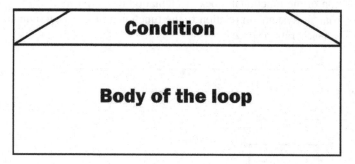

Fig. 3. Semantic element "Loop" of the augmented notation.

The body of the loop will execute as long as the condition is met. The usage of the semantic element "Loop" avoids the problem of the impossibility of displaying the process of a cyclical nature, as well as the most flexible and logical construction of the on-board computing complex.

Figure 4 shows an example of using the "Cycle" semantic element in the decomposition scheme of the "Laser Altimetry" functional block. The "Cycle" element shows that the subsystem logic has been restored. The process must be repeated until the distance to the underlying surface "**h**" is more than zero.

2.3 Meta-modeling of a Subsystem of Star Sensors by Means of Augmented Notation IDEF0

The semantic element "Data stream merging" was proposed in the process of functional modeling of the star sensors subsystem operation. Standard means of the IDEF0 notation don't allow to describe comprehensively the combined processes. It is difficult to show

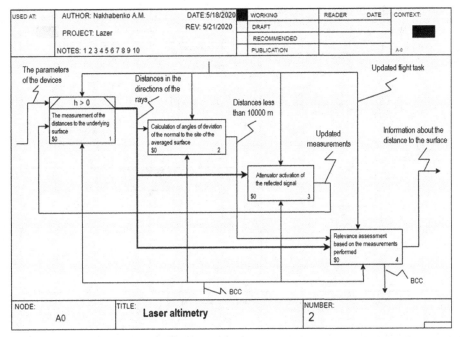

Fig. 4. Decomposition diagram of the functional block "Laser Altimetry" A0 with the semantic element "Loop" in augmented notation.

that the results of the star sensors subsystem and inertial subsystem have the same meaning for the guidance and navigation of the space probe. Figure 5 shows the semantic element "Data stream merging" of the augmented notation.

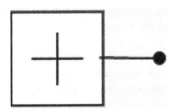

Fig. 5. The semantic element "Data stream merging" of the proposed augmented IDEF0 notation.

The "Data stream merging" element could be used in a case where the functional model of the OBCC becomes more complex. The element is appropriate where some flows after their processes execution should be combined. In this case superfluous arrows only interfere with the reading and understanding of the logic of the processes. Figure 6 shows an example of using the semantic element "Data stream merging" in the decomposition scheme of the functional block "Calculation of landing points using star sensors and inertial navigation".

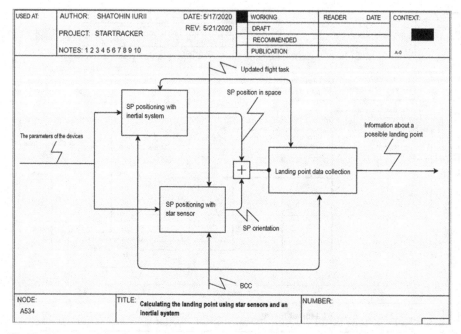

Fig. 6. Diagram of the functional block A534 decomposition "Calculation the landing points using star sensors and inertial navigation system" in augmented IDEF0 notation.

2.4 Meta-modeling of Optical Subsystems of the Direction Finder by Means of Augmented Notation

It was found that it may be difficult if not impossible to describe conditional structures in the process of modeling the optical direction finder subsystem. The standard IDEF0 notation lacks the ability to display conditional structures. It does not allow a diagrammatically description of the relevance between the incoming data and a following function in the model.

The semantic element "Conditional" is intended for organizing the logic of outgoing flows. Thus, for cases when it is necessary to check the result of the process for the fulfillment of a certain condition, the semantic element "Conditional" could be applied (Fig. 7). If the condition is true, a subsystem functional block following the "Yes" are executed. Otherwise, an alternative one will be executed (Fig. 8).

Fig. 7. The semantic element "Conditional" of the augmented notation.

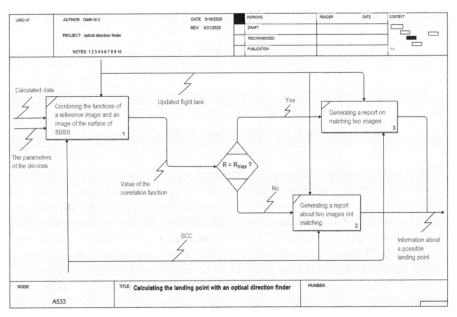

Fig. 8. Diagram of the functional block A533 decomposition "Calculation of the landing point by the optical direction finder subsystem" in the augmented notation.

The semantic element "Conditional" reduces the number of arrows, while maintaining a clarity and intuitively of the schemes and the logic of the on-board computing complex operation.

Figure 8 shows an example of using the "Conditional" semantic element in the decomposition scheme of the "Calculation of the landing point with an optical direction finder" functional block.

2.5 Meta-modeling of a Computer Vision Module Using Augmented Notation

The IDEF0 standard notation has no mechanisms or any other means to reflect the speed of the process and one's time frame. When developing a computer vision subsystem, it is important to have a tool that reflects the limitations on the duration and the speed of the process [22–26]. Based on this, all the operations (image capture, data analysis, etc.) should be described within virtual time constraints. The paper proposes a new semantic element "Chronos" (Fig. 9) for solving the problem. The suggested element allows to make a preliminary calculation of the process execution time and the process execution speed. It is also possible to describe different temporary states of a projected item. The "Cronos" element easily integrates into previously developed functional models.

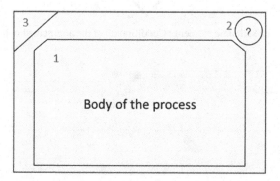

Fig. 9. Semantic element "Chronos".

The semantic element "Chronos" is presented in Fig. 9, where 1 – name of the process; 2 – time limits (for example in seconds); 3 – a presence of decomposition.

The element prescribes an execution of the process inside a block in a limited time. The time limit could affect the result of the execution of both this and the next process. The decomposition diagram of the block "The time limited direction finding to the landing point" using the "Chronos" element in the augmented notation is shown in Fig. 10.

Table 1 provides a general description of all developed semantic elements in accordance with the standards in the field of software engineering and human-machine interaction [27, 28].

Table 1. Description of the developed graphic/semantic elements.

Name of the semantic element	Graphical representation of an element	Description of the semantic element
LCS		The "Local computing complex" element is intended to describe the processes performed using the resources of the information system
Chronos		The "Chronos" element provides a toolkit to initialize the execution time frame of a function block
Loop		The "Loop" element provides a toolkit for describing the multiple repetitions of a function block (loops).
Conditional		The "Conditional" element provides a toolkit to describe a branching of the processes or functional blocks while the system is running.
Data streams merging		The "Data streams merging" element provides a toolkit for mapping the interactions between data streams of two or more processes.

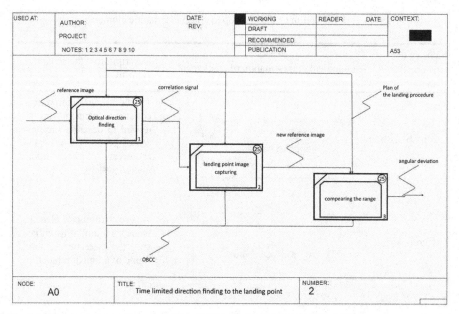

Fig. 10. Decomposition diagram of the block "The time limited direction finding to the landing point" using the "Chronos" element in the augmented notation.

3   Representation of Suggested Semantic Elements in the Form of Programming Code

Nowadays, it is essential to have a means of interpretation (converting) semantic elements of the notation into a programming code. The standard IDEF0 notation does not have this capability.

The presence of an unambiguous relationship "semantic designation – programming code" allows to avoid errors in the program implementation of functional diagrams. It is also a guarantee that the functionality inherent in the model will be implemented. The authors have proposed a mechanism for interpreting augmented notation semantic elements into the programming code. C# was chosen as the programming language.

Table 2 shows a relationship developed in the context of the suggested augmented notation.

Table 2. A single-link dependency "semantic designation – program code".

Semantic designation	Programming code

```
static List<SampleParam<Task>> Process1
(List<SampleParam<Task>> param)
        {
            var result = new
List<SampleParam<Task>>();
            try
            {
                Type m_type =
param.ElementAt(1).m_condition.GetType();
                LocalCC<dynamic> localCC = new Lo-
calCC<dynamic>();
                localCC.instruments = new
List<SampleParam<dynamic>>();
                result = ProcessBody(param, lo-
calCC);
                return result;
            }
            catch (ArgumentException ae)
            {
                throw new Exception("Exception:
"+ae.Source.ToString());
            }
        }
        static List<SampleParam<Task>> Process-
Body(List<SampleParam<Task>> paramList, Lo-
calCC<dynamic> localCC)
        {
            throw new ArgumentNullExcep-
tion("Exception");
        }
```

```
static List<SampleParam<dynamic>> Pro-
cess1(List<SampleParam<dynamic>> param, Sam-
pleParam<double> condition)
        {
            var result = new
List<SampleParam<dynamic>>();
            while (Help-
er.CompareValues(Convert.ToDouble(param.ElementAt(S
ERVICE_FIELD).m_condition), condition.m_condition))
            {
                result = ProcessBodyL(param);
            }
            return result;
        }
        static List<SampleParam<dynamic>> Process-
BodyL(List<SampleParam<dynamic>> paramList)
        {
            throw new ArgumentNullExcep-
tion("Exception");
        }
```

```
static List<SampleParam<Task>> Process1
(List<SampleParam<Task>> paramList, Sam-
pleParam<int> param)
        {
            var result = new
List<SampleParam<Task>>();
            try
            {
                Task task =
Task.Factory.StartNew(() =>
                {
                    result= ProcessBody(paramList);
                });
task.Wait(TimeSpan.FromSeconds(param.m_condition));
                return result;
            }
            catch (AggregateException ae)
            {
                throw ae.InnerException;
```

(continued)

Table 2. (*continued*)

```
        }
    }
        static List<SampleParam<Task>> Process-
Body(List<SampleParam<Task>> paramList)
    {
            throw new ArgumentNullExcep-
tion("Exception");
    }
```

```
    static Tuple<List<SampleParam<dynamic>>, bool> Con-
dition1(List<SampleParam<dynamic>> param, Sam-
pleParam<double> condition)
    {
            var result = new Tuple
<List<SampleParam<dynamic>>, bool>(new
List<SampleParam<dynamic>>(),false);
            if (Help-
er.CompareValues(param.ElementAt(0).m_condition,
condition.m_condition))
                {
                result = new Tuple
<List<SampleParam<dynamic>>, bool>(param, true);
                }
            else
            {
                result = new Tuple
<List<SampleParam<dynamic>>, bool>(param,
false);
            }
            return result;
    }
```

Conditional

```
    static List<SampleParam<dynamic>> MergeThreads
(List<SampleParam<dynamic>> param1,
List<SampleParam<dynamic>> param2)
    {
        try
        {
                if (param1.ElementAt(0).GetType()
== param2.ElementAt(0).GetType())
                {
                    return Enumera-
ble.ToList(param1.Concat(param2));
                }
                else
                {
                    throw new TypeAccessExcep-
tion("Exception");
                }
        }
        catch (ArgumentException ae)
        {
                throw new Exception("Exception: " +
ae.Source.ToString());
        }
    }
```

4 Conclusions

In this paper five semantic elements of the augmented IDEF0 notation were proposed. The semantic element local computing complex (LCS) is necessary to display the computing processes performed by the OBCC subsystems. The semantic element "Chronos" allows correct describing time limited processes. The necessity for the description of the time-limited processes arises when the functional model of the space-probe landing procedure should be built. A large number of OBCC cyclical processes are required to develop and implement the "Loop" semantic element. The semantic element "Conditional" is proposed for the implementation of logical branching of the diagrams. A rigorous description of the branching procedure is essential for a clear treatment and performance of the space-probe OBCC operations. The semantic element "Data streams merging" was developed to ensure a correct description of the OBCC operation. The semantic elements of the suggested augmented notation are developed under the "Do No Harm" principle. All proposed semantic elements could be used without a violation of the original logic of the IDEF0 notation. These elements make it possible to correctly and unambiguously simulate the space probe on-board computing complexes.

One of the most important problems is the providing of the unambiguous understanding of a functional model by both the designer and the developer. The developed rules of interpretation solve this problem. Moreover, the proposed interpretation rules make it possible to automate the process of creating the software part of the space probe OBCC. It is important that the rules for the interpretation of the created semantic elements into the programming code have been developed.

The proposed semantic elements and the rules for their interpretation into the programming code are only one of the steps at modeling the on-board computing complexes of the space probe. It is required to solve some other problems associated with a comprehensive description of the autonomous processes of the OBCC. It seems to be a promising technique to use the modular principle for the OBCC functional model building. The proposed approach allows implementing the algorithm of the functional circuit decomposition for the detailed description of the processes performed by the OBCC. The possibility of designing the space probe OBCC using the suggested augmented notation is also shown.

Acknowledgments. We would like to show our gratitude to Prof. Viatcheslav K. Raev (the Chair of instrumental and applied software of the Russian Technological University) for his useful notes to the draft of the paper. We are also thankful to students Maxim A. Olefir, Aleksey M. Nahabenko and Nikita A. Kuzubov for their assistance with the experimental modeling some processes presented in this paper.

References

1. Low earth orbit (LEO) navigation systems (2020). (in Russian). https://www.kik-sssr.ru/Navigation.htm. Accessed 27 Mar 2020
2. Mikrin, E.A., Mihajlov, M.V.: Satellite navigation of spacecraft in lunar orbit, 27 March 2020. (in Russian). https://www.energia.ru/ktt/archive/2018/02-2018/02-07.pdf
3. Christ, D.R., WernliSr, R.L.: Inertial Navigation System (2020). https://www.sciencedirect.com/topics/engineering/inertial-navigation-system. Accessed 27 Mar 2020
4. Inertial Navigation System Composition (2020). https://www.sbg-systems.com/support/knowledge/what-is-an-inertial-navigation-system. Accessed 29 Mar 2020
5. Aleksander, N., Karol, J., et al.: Inertial Navigation Systems and Its Practical Applications (2020). https://www.intechopen.com/books/new-approach-of-indoor-and-outdoor-localization-systems/inertial-navigation-systems-and-its-practical-applications. Accessed 30 Mar 2020
6. Wen, Zh., et al.: Mathematical Model and Matlab Simulation of Strapdown Inertial Navigation System (2012). https://www.hindawi.com/journals/mse/2012/264537. Accessed 10 Apr 2020
7. Lobanov, A.A., Mordvinov, V.A., Murakov, M.V., Raev V.K.: Building a model of a multifunctional onboard guidance and landing complex for space probe. Softw. Syst. Comput. Methods **2**, 36–50 (2018). https://doi.org/10.7256/2454-0714.2018.2.26217. (in Russian)

8. Portal cfin.ru. https://www.cfin.ru/vernikov/idef/idef0.shtml. Accessed 07 Apr 2020
9. What is a semantic domain. https://semdom.org/description. Accessed 31 Mar 2020
10. Portal britannica.com. https://www.britannica.com/technology/inertial-guidance-system. Accessed 01 Apr 2020
11. Vlasios, T., Catherine, M.: Functional Modeling. https://www.sciencedirect.com/topics/materials-science/functional-modeling. Accessed 02 Apr 2020
12. Portal asqservicequality.org. http://asqservicequality.org/glossary/idef0-integrated-definition-for-function-modeling/. Accessed 02 Apr 2020
13. Portal vitechcorp.com. http://www.vitechcorp.com/resources/core/onlinehelp/desktop/Views/IDEF0.htm. Accessed 03 Apr 2020
14. Description of individual conceptsIDEF0. (in Russian). https://www.cfin.ru/chuvakhin/idef0-r.shtml. Accessed 06 Apr 2020
15. IDEF series notations IDEF. (in Russian). https://bpmsoft.org/idef0-and-idef3/. Accessed 17 Apr 2020
16. Waissi, G.R., Demir, M., et al..: Automation of strategy using IDEF0. https://www.sciencedirect.com/science/article/pii/S2214716015000111. Accessed 21 Apr 2020
17. Methodology of functional modeling IDEF0IDEF0. (in Russian). https://nsu.ru/smk/files/idef.pdf. Accessed 06 May 2020
18. Veis, Š., et al.: Functional and Information Modeling of Production Using IDEF Methods. https://www.sv-jme.eu/?ns_articles_pdf=/ns_articles/files/ojs3/1563/submission/1563-1-1898-1-2-20171103.pdf&id=4931. Accessed 09 Apr 2020
19. Lobanov, A.A., Filonov, A.S.: The method of optical processing of spatial information for the purpose of guidance and landing of space probe on small bodies of the solar system. Cybern. Program. **2**, 94–102 (2018). https://doi.org/10.25136/2306-4196.2018.2.25971. (in Russian)
20. Raev, V.K., Lobanov, A.A.: The method of optical processing of spatial information for the purpose of guidance and landing of spacecraft on small bodies of the solar system. Cybern. Program. **3**(129), 214–217 (2019). (in Russian)
21. Torshina, I.P., Lobanov, A.A., Kuzubov, N.A., Nahabenko, A.M., Olefir, M.V.: Computer vision system as part of the spacecraft guidance and landing onboard complex. Natl. Tech. Sci. **6**(144), 143–146 (2020). (in Russian)
22. Leeuwen, M., Disney, M.: Navigational Sensors. https://www.sciencedirect.com/topics/earth-and-planetary-sciences/inertial-navigation. Accessed 21 Apr 2020
23. Sigov, A., Andrianova, E., Zhukov, D., Zykov, S., Tarasov, I.E.: Quantum informatics: overview of the main achievements. Russ. Technol. J. **7**(1), 5–37 (2019). https://doi.org/10.32362/2500-316X-2019-7-1-5-37. (in Russian)
24. Gecha, V.Y., Zhilenev, M.Yu., Fyodorov, V.B., Khrychev, D.A., Hudak, Yu.I., Shatina, A.V.: Velocity field of image points in satellite imagery of planet's surface. Russ. Technol. J. **8**(1), 97–109 (2020). https://doi.org/10.32362/2500-316X-2020-8-1-97-109. (in Russian)
25. Solarsystem. NASA. Mission Dawn. https://solarsystem.nasa.gov/missions/dawn/overview. Accessed 06 May 2020
26. NASA: Landing the Space Shuttle Orbiter. https://www.nasa.gov/pdf/167415main_LandingatKSC-08.pdf. Accessed 12 May 2020
27. ISO/IEC 25010: Systems and software engineering — Systems and software Quality Requirements and Evaluation (SQuaRE) — System and software quality models (2011)
28. ISO 9241-11: Ergonomics of human-system interaction — Part 11: Usability: Definitions and concepts (2018)

Comparison of Various Algorithms for Scheduling Tasks in a Desktop Grid System Using a ComBos Simulator

Ilya Kurochkin[1](✉) [iD] and Nikolay Kondrashov[2]

[1] Institute for Information Transmission Problems, Russian Academy of Sciences, Bolshoy Karetny Lane 19, build.1, 127051 Moscow, Russia
qurochkin@gmail.com

[2] National Research University Higher School of Economics, Pokrovsky Boulevard 11, 109028 Moscow, Russia

Abstract. A desktop grid system is one of the most common types of distributed systems. The distinctive features of a desktop grid system are the high heterogeneity and unreliability of computing nodes. Desktop grid systems deployed on the BOINC platform are considered. To simulate the functioning of the desktop grid, a modified ComBos simulator based on SimGrid is used. The ComBos simulator adds support for applications with a limited number of tasks, asynchronous execution of multiple applications and various computing resources. Data from existing voluntary distributed computing projects were used to simulate the functioning of the desktop grid. The paper deals with the modification of scheduling system for a desktop grid. Algorithms FS, FCFS, SRPT, and SWRPT were selected from existing heuristic algorithms for comparison. Two heuristic algorithms for scheduling MSF and MPSF tasks were proposed. A simulation of the desktop grid was performed based on data from existing voluntary distributed computing projects. The simulation took into account asynchronous execution of five different computing applications on several types of computing resources. A comparative analysis of the results of various scheduling algorithms in the desktop grid is carried out. Analysis of the results showed that the proposed MPSF algorithm shows the best results from the compared algorithms. The proposed heuristic scheduling algorithm can be applied to umbrella distributed computing projects and to desktop grid in general.

Keywords: Desktop grid · BOINC · Scheduling · ComBos · Umbrella distributed computing project

1 Introduction

With the development of telecommunication networks and the increase in the power of personal computers, distributed systems have become widespread [1]. The desktop grid system (desktop grid) [2] is a type of distributed systems. The desktop grid consists of a server and computing nodes. The server carries out distribution of computational tasks

© Springer Nature Switzerland AG 2020
V. Jordan et al. (Eds.): HPCST 2020, CCIS 1304, pp. 29–40, 2020.
https://doi.org/10.1007/978-3-030-66895-2_3

and aggregation of results. Personal devices such as personal computers, tablets, smartphones and other are mainly used as computing nodes in the desktop grid system. The main distinctive features of the desktop grid include high heterogeneity of computing nodes, the probability of computation errors, unreliability of communication channels and the computing nodes themselves. Computations are often performed in the background on the desktop gird nodes. This allows users to use their personal devices for their intended purpose.

There are software platforms for organizing computations on a desktop grid: BOINC [3], HTCondor, Legion, Globus toolkit, OracleGridEngine, XtremWeb [4], and others. Each platform has its own characteristics when deploying and using a distributed computing system. The most popular platform is BOINC (Berkeley Open Infrastructure for Network Computing). Over 20 years, more than 100 public desktop grids have been deployed on the basis of this platform to solve various scientific problems [5]. Several million personal computing devices were connected to the desktop grid systems based on BOINC [6]. A desktop grid system deployed to solve a single problem is called a distributed computing project. If the computing resources of one or more organizations are used, the project is called an enterprise desktop grid or closed distributed computing project. If computing resources are provided by volunteers, then such a desktop grid is called a voluntary distributed computing project (public desktop grid) [7]. There are additional features and limitations for voluntary distributed computing (VDC) projects. These features and limitations include:

- Fine-tuning of the load balancing system [8].
- Low computational complexity of tasks for the ability to calculate tasks on an average computer for less than 8–9 h.
- Verifying the results.
- Small size of input data and results.
- Problem of long-term calculation of remaining tasks (counting the remains), when the number of nodes is significantly higher than the number of remaining tasks;
- Interaction with the volunteer community.
- Computing application must run on most computing nodes without installing additional software.
- Availability of control points: storing intermediate results on the computing node during task calculation.

This paper discusses how to improve the dynamic scheduling system for issuing tasks. The standard scheme for the BOINC software platform is used when nodes request tasks themselves. As a result, the decision to assign a task to a node can only be made at the time of the request from it. In other words, the task of the algorithm being developed is to make a decision at the time of a request from a node about which task is best to perform on it. The ComBos simulator [9], which is based on SimGrid [10], is used. The ComBos simulator was developed with the purpose of creating a BOINC simulator. This simulator has the following advantages:

1. Supports the use of user data for real projects.
2. Simulates all the services of both the client and server parts of BOINC.

3. Provides ease of development due to high-quality documentation and SimGrid support.
4. Supports good scalability.
5. Provides easy modification of the source code.

The ComBos simulator has the following disadvantages:

6. It does not support umbrella projects and applications with a limited number of tasks;
7. It uses an outdated version of SimGrid.

To solve the problem of dynamic task scheduling, a modified ComBos simulator was used to simulate the server part of the BOINC project. The support for applications with a limited number of tasks, various resources, and umbrella projects (several applications) was added.

2 Scheduling Heuristics

Optimal scheduling in multiprocessor systems is an NP-complete problem [11], which is why many heuristics have been created for it [12]. The main heuristics for scheduling tasks include:

1. Opportunistic Load Balancing (OLB). Each task is assigned to any currently available node. The use of resources is maximized, but the throughput might be very low.
2. Minimum Execution Time (MET). The task is assigned to the node that can calculate it the fastest. Node availability and its current tasks are ignored.
3. Minimum Completion Time (MCT). The same as the MCT, but the time until the node becomes available is taken into account.
4. Min-min. The task with the minimum execution time is given to the node with the minimum execution time for it. This task is removed from the list and the process is repeated until all tasks are distributed.
5. Max-min. The same as min-min, but tasks with the maximum execution time are taken.
6. Duplex. It calculates max-min and min-min and selects the best solution.

The BOINC scheduler uses a modified OLB algorithm by default. When a client requests tasks, the server evaluates which tasks it can calculate on time and sends them. This method allows you to plan one node without any knowledge about the others and only partial information about tasks.

There are also many more complex techniques based, for example, on evolutionary programming [13].

3 Scheduling for Multiple Applications

Most of the papers on scheduling in the desktop grid or Bag of Tasks (BoT) are devoted to the execution of a single set of tasks, rather than its selection from many. For example,

GridBOT [14] almost ignores the problem of selection. They also rarely address the issue of multiple resource types. For example, [15] is an overview of scheduling algorithms in the presence of many types of resources. It examines applications where subsequent tasks depend on previous ones, thus creating a graph of calculations (workflow). The described algorithms are applicable for workflow, not BoT, and are not suitable for scheduling tasks in the desktop grid.

In [16], the problem of choosing a BoT with a known size in the desktop grid is considered, but resource types are not taken into account. Let us describe several scheduling algorithms for running multiple independent applications:

1. Fair share (FS) means that each application is given the same amount of resources per unit of time. It is quite simple to show that such a strategy is not always optimal. Let's say we have two applications that can be executed separately in 1 unit of time. If they both start with FS, both will be completed in 2 units. If you first perform one, and then the second, the first will be performed in one unit, and the second will also be performed in two. This algorithm is used in the World Community Grid umbrella project.

2. First-come-first-served (FCFS) executes applications in the order they are added to the system. This algorithm is indeed more efficient than FS for the same application sizes, but may not be optimal in other cases. For example, if a very long application was added before a short one, the second one will wait for the queue significantly longer than its execution time.

3. Shortest remaining processing time (SPRT) first executes applications with the least amount of work remaining. However, with this strategy, an application can wait indefinitely for its queue if applications shorter than it are constantly added.

4. Shortest remaining processing time (SWPRT) works the same way as SRPT, except that the remaining work is multiplied by some coefficient that depends on the application.

The given list of algorithms is not complete, there are other algorithms. However, they also rely on the fact that the available resources are the same for all applications.

It is also worth mentioning [17], the authors of which study the choice of tasks in the desktop grid systems without additional knowledge about the power of nodes, the total volume of calculations, and so on. The authors of [17] consider only "long" applications and do not take into account the problem of many types of resources. The tasks are similar to the one discussed in this paper, but have different input data. It should be noted that FCFS is divided into 2 variants – FCFS-share and FCFS-exclusive. For the FCFS-share option, applications that don't have tasks in the queue skip ahead to other applications. This does not happen for the FCFS-exclusive option. Only FCFS-share will be considered next.

4   Selecting the Target Function

The target function is introduced to compare different scheduling algorithms and evaluate the performance of an individual algorithm. The target function should be selected taking into account the presence of several independent applications in the grid. Several target functions were considered:

1. Throughput – the number of completed tasks per unit of time

$$\frac{total\_tasks_i}{finish\_time_i - start\_time_i}.$$

   where $total\_tasks_i$ is the number of tasks, and $finish\_time_i$ is the completion time of application i).

2. The execution time (makespan) of a single application

$$finish\_time - start\_time.$$

3. Stretch is the ratio of how long the application was executed to how long it would have been executed if this application was the only one in the system (let's define this time as $perfect\_time_i$ for application i). It can be set using the formula

$$\frac{finish\_time_i - start\_time_i}{perfect\_time_i}.$$

The throughput of different applications is not a good indicator, as it depends on the size of tasks, which may be different. In the case of runtime optimization, large applications will have more weight, because they are initially higher. Stretching is, in a sense, a weighted execution time and is most suitable for this task. It is stretch that is used as a quality metric in the GridBOT project [14].

As part of this work, the maximum stretch will be minimized, since it shows the worst case and avoids application starvation. In other words, the objective function has the following form:

$$\max_{i \in A} \frac{finish\_time_i - start\_time_i}{perfect\_time_i}.$$

5 Selecting Algorithms

FS, FCFS, SRPT, and SWRPT were selected from existing algorithms for comparison. They were taken from the work [16], which also uses stretching as an objective function. The weight for an application in SWRPT is the inverse of the total performance of all the resources on which it can be executed.

In the notations given earlier, these algorithms can be defined as selecting the application for which the minimum value is as follows:

1. FS:

$$\frac{done\_work_i}{t - start\_time_i}.$$

2. FCFS: $start\_time_i$;
3. SRPT: $total\_work_i - done\_work_i$;
4. SWRPT:

$$\frac{total\_work_i - done\_work_i}{total\_power_i},$$

where $total\_power_i = \sum power_j, j \in H, platform_j \in platforms_i$.

Based on the fact that the maximum stretch is minimized, two more algorithms are proposed, based on the assumption that if you choose applications with the maximum stretch at the moment, the value of the objective function will eventually be lower:

1. MSF (maximum stretch first) – selects the application that has the maximum stretch of the already completed part, i.e.

$$\frac{perfect\_time\_done_i}{t - start\_time_i}.$$

Here $perfect\_time\_done_i$ is the time it would take to execute $done\_work_i$, if application i was the only one on the system. In reality, it is not always impossible to find out this time, so we estimate it as

$$\frac{done\_work_i}{total\_power_i}$$

2. MPSF (maximum predicted stretch first) - selects the application that will have the maximum stretch if it uses all the resources available to it before it finishes. In other words,

$$\frac{perfect\_time_i}{t + perfect\_time\_left_i - start\_time},$$

where $perfect\_time\_left_i$ - is the time that the remaining part of the application will run while using all available resources. It is evaluated as

$$\frac{total\_work_i - done\_work_i}{total\_power_i}.$$

MPSF, unlike MSF, takes into account the size of the entire application and the resources available to it, which is why better results are expected from it.

6 Simulating

Before conducting experiments, certain general settings were selected for them. Client records from the SETI@home project [18] for March 2016 were used as client parameters. The experiments did not consider the interaction of multiple projects and network features. In all scenarios, only one project and one user group are considered. Only one data server and one scheduling server are considered, and all task files will be generated on the server side.

The number of errors and incorrect results was determined as 5%. The task replication value is defined as 3. Since a minimum of 2 matching results are required when checking the correctness of the task execution, and if the error level is 5%, in most cases 3 results are sufficient. This choice is due to the fact that a large replication or error rate only increases the time of experiments, increasing the number of copies of tasks that need to be performed.

The configurations used in the experiments differ only in the number of applications, a set of resource types, and parameters: start time, computational complexity of tasks performed, and the number of tasks for applications.

The first experiment: a set of identical applications that are added to the system evenly, and only one type of resource that is used by 1000 clients. Let's take 5 applications with the parameters of computational complexity of tasks $task\_fpops = 3 * 10^{12}$ and the number of tasks $ntotal\_tasks = 120,000$. They will run sequentially with an interval of 100,000 s. This time is not enough to complete a single application, so their execution will overlap.

In this configuration, SWRPT and SRPT work the same way, since the number of available resources for all applications is equal. SRPT doesn't differ from FCFS either, because the application that started earlier will have less work left. The results of the scheduling algorithms are presented in Table 1.

Table 1. Results of the 1st experiment.

Algorithm	Max stretch	Mean stretch
FS	5.04	4.72
FCFS	4.31	2.86
MSF	5.52	4.95
MPSF	4.21	3.24

MSF in this case turned out to be the worst algorithm for both average and maximum stretching. FS performed about 10% better. FCFS and MPSF gave similar indicators – the maximum stretch is almost the same, but the average value for SRPT is 12% lower.

The second experiment involves using 4 different types of resources, with 250 clients for each of them. The list of available types for each application are presented in Table 2, other parameters have remained the same. This distribution is designed to test whether the algorithms will be able to give higher priority to applications 2, 4, because they have fewer resources available. The results of this scenario are shown in Table 3.

FS was significantly worse than other algorithms that give comparable results. It is worth noting that in this case SRPT is better than FCFS, and adding accounting for available resources really improves the result. MSF, unlike the previous experiment, was

Table 2. Available resource types in the 2nd experiment.

Number of application	Available resources
1	0, 1, 2, 3
2	0
3	0, 1, 2, 3
4	1
5	1, 2

Table 3. Results of the 2nd experiment.

Algorithm	Max stretch	Mean stretch
FCFS	2.70	1.86
FS	4.89	2.88
SRPT	2.47	1.77
SWRPT	2.33	1.76
MSF	2.51	2.10
MPSF	2.13	1.78

better than FCFS, but still worse than SRPT and SWRPT. The lowest maximum stretch is again in MPSF.

In the third experiment, various computational applications are taken. Their parameters are shown in Table 4. In this case, there are two long appendices at the beginning, then two short ones, and another long one. This configuration allows you to see how the algorithms cope with the selection of priorities in the case of shorter or longer candidates.

Table 4. Application settings for the 3rd experiment.

Number of application	Start time (seconds)	Number of tasks	Task size (GFLOPs)
1	0	290000	3000
2	10000	120000	3000
3	100000	160000	1000
4	150000	160000	1000
5	200000	120000	3000

To begin with, we will again conduct an experiment with one type of resource (also without SWRPT). The results of this experiment are presented in Table 5 (Fig. 1).

Table 5. Results of the 3rd experiment.

Algorithm	Max stretch	Mean stretch
FCFS	10.67	6.70
FS	5.61	4.59
SRPT	2.81	2.46
MSF	5.51	4.51
MPSF	2.90	2.71

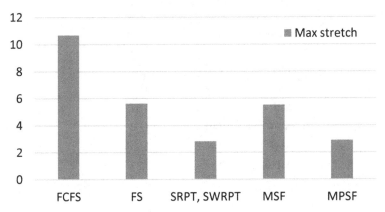

Fig. 1. Results of the 3rd experiment. Maximum stretch.

In this experiment, FCFS turned out to be the worst option – the maximum stretch is more than twice that of any other algorithm. FS and MSF were close in both maximum and average values, but also twice as bad as the best option. As in the case of identical applications, SRPT and MPSF have very similar results, only in this experiment, SRPT has a slightly better value of the objective function.

In the fourth experiment, 4 resource types were added (250 nodes each). The correspondence of applications to types is shown in Table 6. One long application has only one type, another one has two, and one short application has an incomplete set. This experiment tests how well the algorithms will work when there are both long and short applications with fewer resources. The results are shown in Table 7 (Fig. 2).

FCFS and FS in this experiment were also worse than other algorithms. SRPT, compared to the case with a single platform, showed a much higher maximum stretch. MSF was only 18% worse than the best algorithm. SWRPT and MPSF were again close in results.

Several conclusions can be drawn from the experiments:

- Taking into account available resources really improves the result. If you have multiple platforms, it is always better to use SWRPT than SRPT.

Table 6. Available resource types in the 4th experiment.

Number of application	Available resources
1	0, 1, 2, 3
2	0, 1, 2
3	0
4	0, 1, 2, 3
5	1

Table 7. Results of the 4th experiment.

Algorithm	Max stretch	Mean stretch
FCFS	7.62	3.52
FS	5.38	2.59
SRPT	5.04	2.32
SWRPT	2.91	1.90
MSF	3.34	2.79
MPSF	2.83	2.33

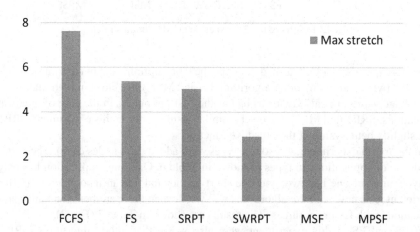

Fig. 2. Results of the 4th experiment. Maximum stretch.

- MSF shows results similar to FS, which are worse than SWRPT and MPSF in all cases.
- MPSF gives good results in all experiments.

The values of the objective function in SWRPT and MPSF are almost identical, but when using SWRPT, there are problems with idle of computing nodes.

7 Conclusion

A comparative analysis of various algorithms for selecting sets of independent tasks in the desktop grid systems is carried out. From the existing algorithms, the FS, FCFS, SRPT, and SWRPT heuristics were selected for comparison. The weight for an application in SWRPT is the inverse of the total performance of all the resources on which it can be executed. Experiments were conducted that showed that the proposed MPSF algorithm shows the best results from the selected options. The results obtained are applicable not only in umbrella projects, but also for execution at the desktop grid as a whole. The modified scheduling system can be used in the existing BOINC projects, and it can help speed up the execution of computational experiments.

Acknowledgements. This work was funded by RFBR according to the research projects No. 18-29-03264 and No. 19-07-00802.

References

1. Foster, I., Kesselman, C.: The grid 2: blueprint for a new computing infrastructure (2004)
2. Choi, S.J., et al.: Characterizing and classifying desktop grid. In: Proceedings - Seventh IEEE International Symposium on Cluster Computing and the Grid, CCGrid 2007 (2007). https://doi.org/10.1109/ccgrid.2007.31
3. Anderson, D.P.: BOINC: a platform for volunteer computing. J. Grid Comput. **18**(1), 99–122 (2019). https://doi.org/10.1007/s10723-019-09497-9
4. Cappello, F., et al.: Computing on large-scale distributed systems: XtremWeb architecture, programming models, security, tests and convergence with grid. Future Gener. Comput. Syst. (2005). https://doi.org/10.1016/j.future.2004.04.011
5. Choosing BOINC projects. https://boinc.berkeley.edu/projects.php. Accessed 15 Sept 2020
6. Project stats info. BOINCstats. https://boincstats.com/en/stats/projectStatsInfo. Accessed 15 Sept 2020
7. Sarmenta, L.F.G., Hirano, S.: Bayanihan: building and studying web-based volunteer computing systems using Java. Future Gener. Comput. Syst. (1999). https://doi.org/10.1016/S0167-739X(99)00018-7
8. Oprescu, A.M., Kielmann, T.: Bag-of-tasks scheduling under budget constraints. In: Proceedings - 2nd IEEE International Conference on Cloud Computing Technology and Science, CloudCom 2010 (2010). https://doi.org/10.1109/cloudcom.2010.32
9. Alonso-Monsalve, S., García-Carballeira, F., Calderón, A.: ComBos: a complete simulator of volunteer computing and desktop grids. Simul. Model. Pract. Theory (2017). https://doi.org/10.1016/j.simpat.2017.06.002
10. Legrand, A., Marchal, L., Casanova, H.: Scheduling distributed applications: the SimGrid simulation framework. In: Proceedings - CCGrid 2003: 3rd IEEE/ACM International Symposium on Cluster Computing and the Grid (2003). https://doi.org/10.1109/ccgrid.2003.1199362
11. Hartmanis, J.: Computers and Intractability: a guide to the theory of np-completeness (Michael R. Garey and David S. Johnson). SIAM Rev. (1982). https://doi.org/10.1137/1024022

12. Braun, T.D., et al.: A comparison of eleven static heuristics for mapping a class of independent tasks onto heterogeneous distributed computing systems. J. Parallel Distrib. Comput. (2001). https://doi.org/10.1006/jpdc.2000.1714

13. Estrada, T., Flores, D.A., Taufer, M., Teller, P.J., Kerstens, A., Anderson, D.P.: The effectiveness of threshold-based scheduling policies in BOINC projects. In: e-Science 2006 - Second IEEE International Conference on e-Science and Grid Computing (2006). https://doi.org/10.1109/e-science.2006.261172

14. Silberstein, M., Sharov, A., Geiger, D., Schuster, A.: GridBot: execution of bags of tasks in multiple grids. In: Proceedings of the Conference on High Performance Computing Networking, Storage and Analysis, SC 2009 (2009). https://doi.org/10.1145/1654059.1654071

15. George Amalarethinam, D.I., Maria Josphin, A.: Dynamic task scheduling methods in heterogeneous systems: a survey. Int. J. Comput. Appl. (2015). https://doi.org/10.5120/19318-0859

16. Legrand, A., Su, A., Vivien, F.: Minimizing the stretch when scheduling flows of divisible requests. J. Sched. (2008). https://doi.org/10.1007/s10951-008-0078-4

17. Anglano, C., Canonico, M.: Scheduling algorithms for multiple bag-of-task applications on Desktop Grids: A knowledge-free approach. In: IPDPS Miami 2008 - Proceedings of the 22nd IEEE International Parallel and Distributed Processing Symposium, Program and CD-ROM (2008). https://doi.org/10.1109/ipdps.2008.4536445

18. Korpela, E., Werthimer, D., Anderson, D., Cobb, J., Lebofsky, M.: SETI@HOME - Massively distributed computing for SETI. Comput. Sci. Eng. (2001). https://doi.org/10.1109/5992.895191

Information Technologies and Computer Simulation of Physical Phenomena

Thermal and Microstructural Analysis of Intermetallide Synthesis in the Ni-Al Layered-Block Atomic Structure Based on the Computer-Aided Simulation of SHS

Vladimir Jordan[1,2]([envelope]) [ORCID] and Igor Shmakov[1] [ORCID]

[1] Altai State University, Lenin Avenue 61, 656049 Barnaul, Russia
jordan@phys.asu.ru
[2] Khristianovich Institute of Theoretical and Applied Mechanics of SB RAS, Institutskaya Street 4/1, 630090 Novosibirsk, Russia

Abstract. The paper presents the results of thermal and microstructural analysis of SH-synthesis of intermetallic compounds in the layered-block atomic structure of Ni-Al based on computational experiments on computer simulation of the SHS process. The layered block structure consists of five superimposed elongated layers. Each layer contains 6 pairs of blocks of two different types (i.e. 12 blocks). The block of the first type is a crystal lattice of Ni atoms and the block of the second type is a crystal lattice of Al atoms. In odd layers, pairs of blocks (Ni, Al) follow one after another. In even layers, pairs follow each other in the transposition form (Al, Ni). Computational experiments were carried out on a cluster of 15 PCs using the LAMMPS package in the version of parallel computing based on the method of molecular dynamics simulation. Among the computational experiments results, one should note the "pulsating (oscillatory)" mode of propagation of the SHS combustion wave, which is confirmed by the calculated sets of temperature and density profiles of the substance.

Keywords: SH-synthesis · Molecular dynamics method · Elementary crystalline cell · Temperature profile · Substance density profile · Parallel computing · LAMMPS and OVITO packages

1 Introduction

Nickel-based alloys and superalloys are characterized by high corrosion resistance at high temperatures and they have found wide application in modern mechanical engineering [1]. For example, the share of such superalloys of the total mass of an aircraft engine traditionally reaches 50%. Nickel alloys, including nickel aluminides, are widely used in turbines and combustion sections of an engine, being subjected to significant thermal shock loads during operation. In recent decades, intermetallic compounds $NiAl$, Ni_3Al, etc. have been obtained using the technology of self-propagating high-temperature synthesis (SHS).

© Springer Nature Switzerland AG 2020
V. Jordan et al. (Eds.): HPCST 2020, CCIS 1304, pp. 43–61, 2020.
https://doi.org/10.1007/978-3-030-66895-2_4

SHS as a method for creating new compounds and composite materials with unique properties and characteristics is a process of propagation of a combustion wave through a mixture of reagents (as result of exothermic chemical reaction) with a large release of thermal energy. SHS-materials are synthesized in the form of solid final products [2]. A localized thin layer of exothermic heat release defines the concept of a "front" of a combustion wave, and the combustion wave moves from layer to layer. The initiation of the "ignition" process of a reagent mixture is usually carried out on its surface by the action of a short-term heat pulse lasting up to one second. For example, the "ignition" process is usually carried out by the action of an electric spiral, an electric spark discharge, a laser beam, etc. In the SHS reaction, reagents are used in the form of fine powders (compacted agglomerates or bulk density powders), thin films, liquids, and gases. The combustion wave is characterized by a certain length and is divided into a number of zones [2]:

- heating zone or, in other words, the zone adjacent to the reaction zone (heat transfer and heating of the mixture of reagents occur, but combustion is still absent);
- reaction zone (there is the combustion reaction zone with exothermic heat release);
- afterburning zone or aftercombustion zone (chemical reactions continue in it, which practically do not affect the speed of the combustion wave front);
- zone of secondary physical and chemical transformations, which determine the composition and structure of the final products of SHS.

The propagation of the combustion wave in the volume of the mixture of reagents is considered the first (main) stage of SHS. Simplified, the SHS process can be represented in the form of a structural diagram (process formula, [2]):

$$SHS = Combustion + Structure\,Formation.$$

The second stage of SHS is determined by secondary physicochemical transformations. The composition and structure of final products with a variety of micro-, meso- and macrostructures of various scales of heterogeneity are influenced by the following factors: porosity and temperature of the reagent mixture, their dispersion and stoichiometric initial ratio, dilution degree of the mixture with inert additives, heat loss to the environment, stability of the movement of combustion wave front, etc.

The mutual influence of elementary processes occurring in the reaction medium at different levels of the hierarchy of its heterogeneity forms the behavior of the combustion wave at the macroscopic level. And the microheterogeneous structure of the combustion wave is displayed in the combustion "micro-foci" structure (in the so-called "discreteness problem" of SHS) and largely determines the mode of discrete decay of the thermal structure of the combustion wave. In other words, the discrete decay of the thermal structure of the combustion wave characterizes a certain degree of instability of the combustion wave.

The study of combustion microkinetics, phase and structure formation in the SHS process during physical experiments requires significant financial, material and time costs, thereby seriously complicating the development of optimal control modes for the SHS process. The computer simulation methods of the SHS process make it possible to perform a much larger number of computational experiments (CEs) in comparison with physical experiments. Additionally, the CEs allow such research to perform in a more targeted and predictable manner.

2   Using the LAMMPS Package in a Parallel Computing Configuration for SHS Simulation in a Layered-Block Ni-Al System

For atomic systems containing up to several million atoms in their composition, a fairly effective method for studying the evolution of a system under conditions of temperature-force action on it is the "molecular dynamics (MD)" method. One of the most effective versions for software implementation of the MD method is the LAMMPS software package.

2.1   Summary of MD Method, EAM Potential and LAMMPS Package

The MD method is closely related to statistical physics, in which the atomic structure (system of atoms) acts as a mechanical system of N bodies (particles) interacting with each other, as a result of which the spatial coordinates of atoms change in time, and consequently their velocities, energy, temperature and pressure change in the atomic system. That is, along with a change in the microscopic parameters of the particles, the macroscopic parameters of the atomic system as a whole also change. However, for an atomic system, one can consider such equilibrium states (conditions or restrictions) in which some of the macroparameters of the system retain their values. In other words, these constraints are called ensembles of thermodynamics in statistical mechanics.

For example, for the canonical ensemble NVT and the corresponding Nose-Hoover thermostat, the following values are retained: the number of atoms N, the volume V and the temperature T. The microcanonical ensemble NVE, where E is the internal energy of the system, and the corresponding Langevin thermostat can be used to quickly achieve a certain equilibrium state of the system. The evolution of the system can be monitored with the NVE ensemble without requiring a thermostat condition. Another canonical ensemble is the NPT ensemble with thermostat and barostat, where P is pressure.

The computational procedure of the MD method calculates the trajectories of particles by integrating the equations of motion (Newton's equations), for example, using the Verlet multi-step algorithm [3, 4], taking into account the discrete time step Δt:

$$
\begin{cases}
v_i(t + \Delta t/2) = v_i(t) + \dfrac{F_i(t)}{2m_i}\Delta t, \\[2mm]
r_i(t + \Delta t) = r_i(t) + v_i(t + \Delta t/2)\Delta t, \\[2mm]
F_i(t + \Delta t) = -\nabla_i U(r_i(t + \Delta t)), \\[2mm]
v_i(t + \Delta t) = v_i(t + \Delta t/2) + \dfrac{F_i(t + \Delta t)}{2m_i}\Delta t,
\end{cases}
\tag{1}
$$

where $v_i(t)$ is the speed of the i-th atom; $r_i(t)$ is radius vector of the i-th atom; $F_i(t)$ is vector force acting on the i-th atom; m_i is the mass of the i-th atom; $\nabla_i U(r_i(t))$ is the gradient from the potential $U(r_i(t))$, which determines the force $F_i(t)$ acting on the i-th atom. In the first and last formulas of the system of Eqs. (1), the ratios of the values of the force $F_i(t)$ and $F_i(t + \Delta t)$ to the mass m_i determine the values of the accelerations of the i-th atom at times t and $t + \Delta t$, respectively. The force $F_i(t)$ acting on the i-th atom is determined using the gradient from the multiparticle potential $U(r_i(t))$ as the total result of interaction with each "neighboring" atom (the number of neighboring atoms is determined by the so-called "cutoff" radius). As mentioned above, having at our disposal sets of coordinates and velocities for all atoms of the system at successive times, it is possible to calculate the energy of the system of atoms, temperature, and other parameters of the system at the same times. The accuracy of such calculations depends on the choice of the model function $U(r)$, which is the interatomic interaction potential.

Under conditions of temperature-force effects on atoms in the process of propagation of an exothermic combustion wave of SHS, high-pressure bursts, internal defects and dislocations appear in the local structures of the atomic system. Therefore, stringent requirements are imposed on the accuracy of the multiparticle potential of interatomic interaction in terms of simulation the energy and structural transformations occurring during the propagation of the SHS combustion wave through the reagent mixture.

The authors of the publication [5] proposed the "embedded atom model (EAM)" for the Ni-Al atomic system as a universal potential of interatomic interaction. The EAM potential is generally equally well suited for the intermetallic phases NiAl, Ni_3Al and for pure Ni and Al. The EAM potential table file available for download (http://www.ctcms.nist.gov/potentials/), intended for building the initial Ni-Al atomic system, can be used as an input file for the LAMMPS software package [6] (software available at http://lammps.sandia.gov/). The graphs of the functions of pair interactions of the EAM potential [5], published in 2009 for the Ni-Al system, are shown in Fig. 1.

EAM potentials, used in combination with the MD method, are quite effective and accurate in simulation and evaluating the mechanical, thermal and structural properties of metallic systems.

In this work, the free software package LAMMPS is used - one of the most effective versions of the software implementation of the molecular dynamics method, which allows users to perform simulations for various types of particle systems (up to biosystems) with a large number of atoms (up to tens of millions of atoms) with using a wide range of potentials. The LAMMPS version, written in C++ with the MPI message passing interface, allows parallel computations and significantly accelerates a large amount of computations when simulating the processes under study [6]. The package is distributed under the GPL license and is available as source code. Before starting the simulation in the LAMMPS package, using the input file, the atomic system is built, the potentials are assigned and the simulation procedures are set up.

Let's pay attention to the main advantages and features of the LAMMPS package [6]:

– the package allows you to solve Newton's equations of motion for systems consisting of atoms, molecules or macroscopic particles that interact with the help of short- or long-range forces with different initial and/or boundary conditions;
– the software implemented most of the many-particle short-range potentials of interatomic interaction, including and two-particle potentials;

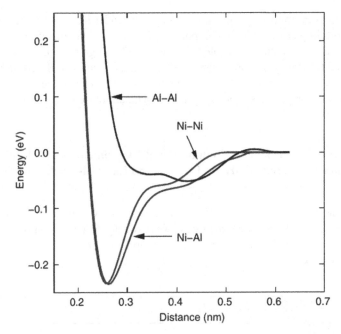

Fig. 1. The graphs of the functions of pair interactions of the EAM potential [5].

- in systems in which the forces of the Coulomb interaction are calculated, the software-implemented methods of Ewald and PPPM can be used;
- for the system under study, it is possible to analyze the current state of the atomic configuration, which can be unloaded both into text and binary file. The initial configuration of atoms used for further calculations can be generated both in the program and loaded from a binary/text file;
- if a large amount of calculations is required, then LAMMPS is compiled in the parallel computing version, but it can be configured to run in a single-processor mode;
- parallel computations are performed on multiprocessor systems using MPI, and the CUDA technology using GPU processors, unfortunately, can only be used for the Lennard-Jones and Coulomb potentials. For this purpose, spatial decomposition methods are used to divide a "3D-simulation domain", for example, a 3D-parallelepiped, into small 3D-subdomains, each of which is assigned to an independent (separate) processor.

LAMMPS uses a "neighbor list model" for particles to keep track of nearby particles and thus provide high computational efficiency. The lists are optimized for systems with particles that repel at short distances, so that the local particle density never gets too high and the particles fill the 3D-parallelepiped with approximately the same density. Processors exchange information and store information about the atoms contained in their subdomain for those atoms that are in the neighboring subdomains.

To construct images with the distributions of atoms and types of structures corresponding to unit cells (fcc, bcc, hcp, etc.), the OVITO program is used [7, 8], which also uses spatial decomposition into subdomains. For rendering you can use several ray tracing methods implemented in the components: OpenGL, Tachyon, OSPRay and POV-Ray, which can be enabled (or disabled) in the OVITO program configuration. In this work, we used the Tachyon component based on the Tachyon Parallel/Multiprocessor Ray Tracing System. This type of ray tracing allows you to use all the threads that can be generated in a computing system (a computer with a multi-core processor) to render the final image.

2.2 Technique for Performing Computational Experiments on Molecular-Dynamics Simulation of SHS in Layered and Layered-Block Atomic Systems Ni-Al

As an example of a layered nanoscale atomic system Ni-Al, Fig. 2 shows a structure for which the results of simulation the mode of microheterogeneous combustion during SHS were obtained in [9] based on a cycle of computational experiments (CEs) using the LAMMPS package.

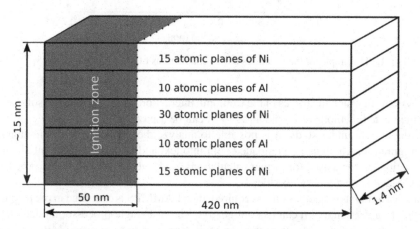

Fig. 2. The structure of a layered nanosized atomic system Ni-Al [9].

According to the technique for performing computational experiments (CEs) [9], each of the five layers of the Ni-Al atomic system (Fig. 2) was a package of a certain number of atomic planes in the form of a crystal lattice with unit cubic cells of the fcc-type (both for a layer of Ni atoms and layer of Al atoms). The values of the parameter a of the cubic cells of the Ni and Al metals differ: $a = 0.3524$ nm for Ni, and $a = 0.405$ nm for Al [5, 10]. The stoichiometric ratio of the components was expressed as the ratio of the number of Ni atoms to the number of Al atoms, which in this case was defined as $N_{Ni}/N_{Al} = 3.94$. In other words, if we recalculate the relative fractions, then the fraction of Ni atoms was $n = 0.7975$ (79.75%), and the fraction of Al atoms was 0.2025 (20.25%). The number of all N atoms in the structure was 717410. At the beginning of the computational experiment (CE), the sample under study in its entire volume (Fig. 2) was uniformly heated for 0.4 ns at a temperature of 600 K. In this case, there was a "relaxation" of the structure of the atomic system with fixed thermodynamic parameters (NPT-ensemble): $N = 717410$ – the number of all atoms in the structure, $P = 1$ Bar – external pressure, temperature $T = 600$ K. In addition, for all three measurements for the atomic structure under study, periodic boundary conditions were established, which remained at this stage of the simulation for 0.4 ns.

At the second stage of simulation, for 0.1 ns in the initial zone of the system under study, within 50 nm (as shown in Fig. 2), heating was carried out, starting from 600 to 1200 K, while maintaining periodic boundary conditions for the sample structure. Heating within 50 nm was carried out under the conditions of the NVT ensemble (V is the volume of the heating zone), and in the remaining part of the volume (in the range from 50 to 420 nm) during the same 0.1 ns, the conditions of the NVE ensemble were observed. Value E is the total energy atoms of the Ni-Al system. After 0.1 ns of the second stage, "free" boundary conditions were fixed in the structure along the X direction, and periodic conditions were retained along the Y and Z directions. Thus, by the end of 0.5 ns under the conditions of the NVE-ensemble in the initial zone of the layered structure (Fig. 2), the SHS process was initiated with further propagation of the combustion wave along the direction of the X axis.

In this paper, using the LAMMPS package with support for parallel computing and the EAM potential of 2009 [5], the CEs results are presented for SHS simulation and studying the mode of microheterogeneous combustion in a model atomic "layered-block" structure (LBS) with alternating nanosized blocks of Ni and Al atoms (Fig. 3). As can be seen from Fig. 3, each of the 5 layers contains 12 blocks with a length (width) of 35 nm. The upper and lower layers are equal in height to half the height of each of the 3 inner layers. Thus, the height of the inner layer is approximately 4.125 nm, and the third dimension for the blocks is 2.5 nm. In other words, the Ni-Al system in the LBS form consists of 12 vertical stacks of 5 blocks each. The dimensions of each vertical stack of 5 blocks are determined as $35 \times 2.5 \times 16.5$ nm.

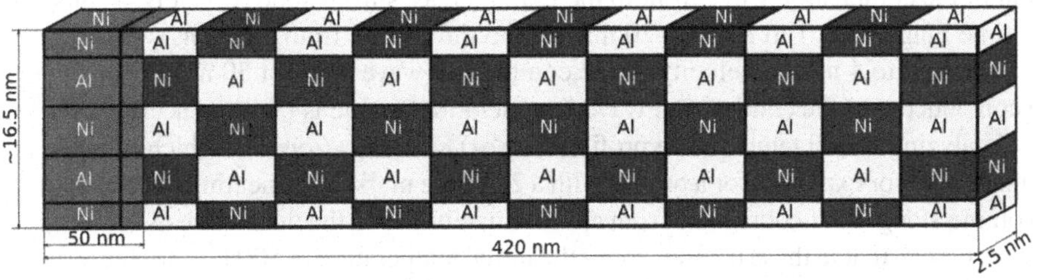

Fig. 3. The structure of a layered-block nanoscale atomic system Ni-Al.

The structure shown in Fig. 3 includes 992340 Ni atoms and 458575 Al atoms, i.e. the total number of atoms $N = 1450915$. Thus, the stoichiometry of the composition of the components is determined by the ratio $N_{Ni}/N_{Al} = 2.164$, and their fractions are distributed as follows: the fraction of Ni atoms is 0.684 (68.4%) and the fraction of Al atoms corresponds to 0.316 (31.6%). In contrast to the layered structure shown in Fig. 2, in the layered-block structure (Fig. 3) the initial heating of the entire volume was carried out for 0.4 ns at the temperature of 800 K. Then, within 0.1 ns within 50 nm of the initial zone of LBS the heating was carried out from 800 to 1400 K. There are no other differences in the methodological conditions and stages of CEs for SHS simulation in LBS (Fig. 3) with respect to the conditions for performing CEs for a layered structure (Fig. 2).

3   Thermal and Microstructural Analysis of Intermetallide Synthesis in the Ni-Al Layered-Block Atomic Structure

To estimate the temperature in a 3D layered-block structure (LBS), the entire volume of LBS is divided into domains (parallelepipeds) along the LBS length equal to 420 nm and the corresponding X coordinate, i.e. in the motion direction of the combustion wave. The height of the domains (direction Z) coincides with the height of the LBS and is equal to 16.5 nm, the length of the domain (direction X) is taken equal to 4 nm (a rather small value). The domain depth (Y direction) coincides with the LBS depth and is equal to 2.5 nm. The sizes of the domains are the same and rather small, and on this basis, the "averaged" temperature value in each LBS domain along the X direction is estimated. The averaged value of the temperature in the domain is determined through the calculated total internal energy of the atoms in the volume of the domain. Thus, let us define the concept of "temperature profile" of the LBS (Fig. 4) as a sequence of temperature values corresponding to a sequence of domains along the X direction (to a sequence of X-coordinates of domain centers). Considering the length (width) of one block, equal to 35 nm, we can say that there are almost 9 domains per block for assessing the temperature within one block (almost 9 points on the temperature profile), and the entire temperature profile contains about 105 points (the length ratio 420 nm to a domain length of 4 nm).

Estimating the linear movement of the combustion wave front and the corresponding time when analyzing temperature profiles (Fig. 4), the estimate of the velocity of movement of the combustion wave front in the LBS increases from approximately 40 m/s (in the time range from 1 to 2 ns) to 70 m/s (in the time range from 4 to 5 ns). In the time range from 2 to 4 ns, the velocity of the combustion wave is about 50 m/s. Thus, it can be concluded that the combustion wave "acceleration" mode is taking place.

Analyzing for all temperature profiles (Fig. 4) «heating» zones in which there is no combustion (for example, for a profile with a 2 ns time marker, the heating zone occupies a volume along the X coordinate from approximately 125 to the final value of 420 nm), it is clearly seen that in these heating zones the initial temperature of 800 K is not preserved, to which the LBS was heated uniformly throughout the volume. As can be seen from Fig. 4, the temperature in the heating zones after 2 ns grows rather quickly and reaches the SHS "ignition" temperature (about 1000 K), since according to the phenomenon of thermal conductivity, a significant amount of thermal energy has entered the heating zone in the previous time. Obviously, after 2–3 ns, the temperature in the heating zones becomes significantly higher than the initial value of 800 K, which corresponds to the same zones in the time range up to 0.5 ns. In addition, in the period from 2 to 3 ns (see Fig. 4), the temperature in the heating zones becomes quite close to the ignition temperature of SHS (the first phase of $NiAl_3$ is formed at 1127 K). And therefore, when

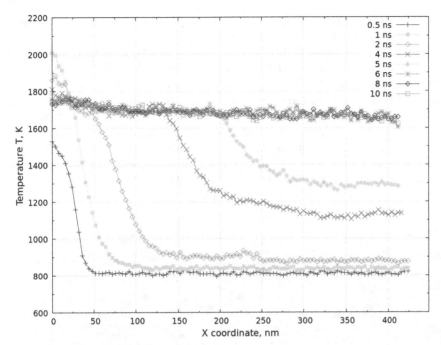

Fig. 4. The temperature profiles corresponding to a sequence of time moments during the propagation of the combustion wave in a layered-block Ni-Al system.

the combustion wave approaches these zones, combustion ignition occurs much faster, contributing to the "acceleration" of the combustion wave front.

For a layered structure (Fig. 2) in publication [9], an almost 2 times lower estimate of the velocity, equal to 25 m/s, was obtained. Almost 2-fold superiority of the velocity of the combustion wave in the LBS in relation to the layered structure is explained by the higher value of the specific surface area of the contact of Ni and Al atoms in the LBS (Fig. 3) compared to the same parameter in the layered structure (Fig. 2). That is, in LBS, the diffusion of atoms spreads not only along the Z axis, perpendicular to the horizontal planes that are the interfaces between layers, but also in each layer along the X axis, perpendicular to the vertical planes separating blocks with different atoms. In a layered system (Fig. 2), diffusion of atoms spreads only along the Z axis.

In addition, additional arguments for this superiority are the higher values of the temperatures of the initial heating of the entire LBS and the subsequent heating of its initial zone (within 50 nm) in comparison with the analogous parameters for the layered structure. The value of the velocity of the combustion wave in powder mixtures with particle sizes of Ni and Al in the range of 10–50 μm turns out to be 2–3 orders of magnitude less than in LBS, and is in the range from about 1 to 20 cm/s. This can be explained by the fact that in real powder mixtures, the specific contact surface of Ni and Al particles, respectively, is also 2–3 orders of magnitude smaller. In confirmation of the results of molecular dynamics simulation of SHS in nano- and microsized atomic

Ni-Al systems it can be noted that physical SHS experiments carried out in thin films (nanofoils, [11]) show high combustion velocity (up to several m/s).

Analysis of Fig. 4 shows that in the layered-block Ni-Al system during the passage of the combustion wave (by the end of 10 ns) a temperature plateau is established with a small decrease from 1700–1720 K at the beginning of the structure to values in the range of 1630–1670 K at the end of structure. The dissolution of the Ni solid phase in the Al liquid phase "in general" (with the exception of some local volumes in these LBS-blocks) is practically completed in 7 ns (the layered-block character of the structure is no longer observed), which is confirmed by the results shown in Figs. 5, 6 and 9. Figures 5 and 6 (taking into account the LBS length of 420 nm) also confirm the acceleration of the combustion wave velocity to 70 m/s and higher. The images shown in Figs. 5, 6, 7 and 8 were obtained using the OVITO program [7, 8].

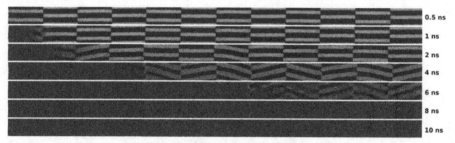

Fig. 5. The sections (images) with the distribution of Ni atoms (blue dots) and Al atoms (yellow dots) over the LBS volume, corresponding to the sequence of time moments during the propagation of the combustion wave). In the case of repainting the colors in grayscale: blue is converted to dark gray, and yellow to light gray. (Color figure online)

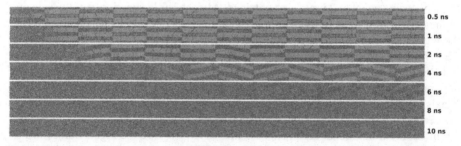

Fig. 6. The sections (images) with the distribution of various types of unit cells (fcc, bcc, hcp, ico, "other") over the LBS volume corresponding to the sequence of time moments during the propagation of the combustion wave: green dots – fcc type; blue dots – bcc type; red dots – hcp type; yellow dots – ico type; white dots – "other" types. In the case of repainting the colors in grayscale: green and yellow are converted to light gray, blue and red to dark gray. (Color figure online)

In accordance with the diagram of the equilibrium state of the Ni-Al system, taking into account the concentration (at.%) of Ni atoms, equal to 68.4%, and at temperatures below the melting point of the Ni₃Al phase, equal to 1653–1658 K, two phases are formed [12, 13]: the NiAl phase and the Ni₃Al phase (with the dominance of the NiAl phase). By the time instant of 6 ns in the second half of the LBS, the temperature plateau corresponds to the range of 1630–1700 K with noticeable temperature "pulsations" (Fig. 4), in which the melting point of the Ni₃Al phase is located. Ni₃Al phase has a superstructure with an fcc-lattice of the L12 type. In local volumes, corresponding within each pulsation to a temperature exceeding the melting point of the Ni₃Al phase (above 1653–1658 K), the Ni₃Al phase decomposes and, as a result, an additional amount of the NiAl phase is formed with an ordered bcc-lattice of the B2 type. The increase in the number of bcc-type structures corresponding to the NiAl phase is confirmed in the table below. On the other hand, in local volumes corresponding within each pulsation to a temperature below the melting point of the Ni₃Al phase, the Ni₃Al phase is also formed. The formation of the Ni₃Al phase will be all the more likely if in these local regions there are substructures of the crystal type, consisting only of Ni atoms (highlighted in blue in Fig. 7).

In Fig. 7, an enlarged image of one of the small fragments is presented, highlighted at the vertical border dividing the 7th and 8th vertical stacks of the LBS (see the 5th thin section in Fig. 5, marker of 6 ns). In fact, the first 7 stacks in 6 ns have already undergone dissolution "in general" (with the exception of some local volumes in the LBS-blocks) and are practically no longer detected (the right edge of the 7th vertical stack is only partially recognized), and the remaining 5 vertical stacks of 12 initial ones are still quite noticeable. In the upper part of Fig. 7 (to the left of the middle of the image), a crystal-type substructure is clearly traced, consisting only of Ni atoms (highlighted in blue) and similar to a "circle (disc)", and around it and in the downward direction, crystal-type substructures from Al atoms (highlighted in yellow). As mentioned above in this paper, both lattices of Ni atoms and lattices of Al atoms correspond to the same fcc-type. In the direction "downward" from the "disc", parallel to the crystalline substructure of Al atoms, there is a crystalline substructure with alternating Ni and Al atoms (highly likely, a crystal bcc-lattice of the B2 type corresponding to the NiAl phase).

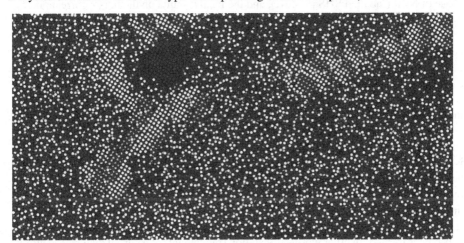

Fig. 7. An enlarged image of one of the small fragments highlighted on the vertical border dividing the 7th and 8th vertical stacks of LBS (see the 5th section in Fig. 5, marker of 6 ns). (Color figure online)

Near the crystalline substructure of Ni atoms in the form of a "disc" and in the crystalline substructures of Al atoms (downward from the "disc"), a chain of reactions (2) rapidly occurs in the stage of intense combustion (combustion temperature 1630–1700 K) with a stable the formation of the NiAl phase. The chain of reactions (2) is accepted by many authors of publications, including and the authors of publications [12, 13]:

$$Ni + 3Al \rightarrow NiAl_3, \quad Ni + NiAl_3 \rightarrow Ni_2Al_3, \quad Ni + Ni_2Al_3 \rightarrow 3NiAl. \qquad (2)$$

Due to the remaining reagent Ni, a stable Ni_3Al phase is formed in those local regions in which the temperature is below the melting point of the Ni_3Al phase (below 1653–1658 K).

In addition, a stripe (layer) of the image with a predominance of crystalline substructures of Al atoms is visible in the upper right fragment in Fig. 7. This is a large part of the upper layer with a lattice of Al atoms from the 8th vertical stack (see the 5th thin section of Fig. 5, marker of 6 ns). In this layer (Fig. 7), as well as in other similar layers of the 8th vertical stack (see the 5th thin section of Fig. 5, marker of 6 ns), transverse parallel narrow crystalline interlayers are clearly visible - substructures with alternating Ni and Al (highly likely, bcc-lattices of the B2 type, corresponding to the NiAl phase) separating such layers of Al atoms with approximately the same repetition period. Crystalline substructures of Al atoms remain between the interlayers with the assumed NiAl phase. The X-coordinates of the temperature minima on the temperature profile (Fig. 4, marker of 6 ns) are comparable to the X-coordinates of the vertical boundaries of vertical stacks (from the 7th to the 12th vertical stack inclusive). However, between the minimums of the temperature profile, corresponding to the boundaries of vertical stacks, the pulsation bursts show several local minima and maxima, which correspond to alternating transverse interlayers in the Al layers, considered a little above. Thus, the local maxima on the bursts of the temperature profile correspond to the interlayers of the high-temperature phase of NiAl (the final SHS product), and the local minima on the bursts correspond to transverse interlayers with crystalline substructures of Al atoms, in which, highly likely, the first stages of the SHS reaction with the formation of the $NiAl_3$ and Ni_2Al_3 phases occur. The analysis of the presence of noticeable temperature fluctuations at a time point of 6 ns, carried out on the basis of Figs. 4, 5 and 7, is confirmed (albeit less clearly) by a similar arrangement of the above-mentioned substructures in Fig. 8.

In Fig. 8 (similar to Fig. 7), in the upper part to the left of the middle of the image, there is a crystalline substructure of the fcc-type, also similar to a circle (disc) and which corresponds to the lattice of Ni atoms, colored green. However, in the crystalline substructure of Ni atoms (in Fig. 8 along the "diameter of the circle"), a stacking fault is observed in the form of a separating layer of 2 atomic Ni planes corresponding to the hcp-type ("hexagonal close packing"). In Fig. 8, these two atomic planes are marked in red. It is likely that due to the excess pressure from both sides on the substructure along the perpendicular direction to the "diameter of the circle" and the sliding of atomic planes relative to each other along the "diameter of the circle", a layer of Ni atoms with the hcp-type is formed. In other words, translation (movement) of a partial dislocation along the slip plane in the fcc-structure of Ni atoms (in the circular substructure corresponding to Figs. 7 and 8) leads to the appearance of a narrow layer of the hcp-structure of Ni

Fig. 8. An enlarged image of one of the small fragments highlighted on the vertical border dividing the 7th and 8th vertical stacks of LBS (see the 5th section in Fig. 6, marker of 6 ns). (Color figure online)

atoms. In fact, quite a few similar hcp-substructures can be found in other fragments of Fig. 8, and in the corresponding fragments of Fig. 7 one can observe "clusters" of Ni atoms - agglomerations of Ni atoms (blue dots).

Continuing the analysis of Fig. 8, a layer with alternating transverse crystalline interlayers of 2 types is also noticeable in the upper right fragment of it, although less distinctly than in a similar fragment of Fig. 7. In the transverse interlayers of one of the types, sublattices of atoms are distinguished, painted in blue. The OVITO program marks in blue the atoms that make up bcc-structures. I.e., the obtained result of the analysis of Fig. 8 confirms the assumption formulated in the analysis of Fig. 7 that the corresponding transverse interlayers with alternating Ni and Al atoms are bcc-lattices of the B2 type, in other words, correspond to the intermetallic phase NiAl. In transverse interlayers of another type of the same layer (Fig. 8), adjacent to interlayers of the first type, the OVITO program uses green color to indicate Al atoms (fcc-structures), which are marked in yellow in Fig. 7. Similar conclusions on the recognition of substructures corresponding to the intermetallic phase of NiAl can be drawn for the strip (layer) in the downward direction from the substructure in the form of a "circle (disc)" shown in Figs. 7 and 8.

Returning to an additional explanation of the presence of pulsations on the temperature profiles, we note that as a result of the "acceleration" of the combustion wave, the velocity of the combustion front becomes significant (70 m/s and higher) and by the end of 6 ns the combustion wave front as a whole reaches the end of the LBS (mark 420 nm), bypassing some local volumes, the boundaries of which are associated with the presence of any stacking faults and a sharp transition from one substructure to another. That is, in some local volumes, as shown in Fig. 7, the multistage SHS process has not yet begun, and in the rest of the reaction volume, the process of dissolution of the system components is almost complete and the final products of the SHS reaction (intermetallic phases NiAl and Ni_3Al) are formed. It should be noted that the profiles with markers of 8 and 10 ns also retain temperature fluctuations in the range of 1630–1670 K, although with a lower amplitude compared to the profile corresponding to the 6 ns marker. Temperature pulsations persist, despite the fact that the dissolution process of the components of the initial Ni-Al system by the end of 10 ns was completed "in general", since the layered-block structure in Figs. 5, 6 and 9 is practically not observed and the active stage of SHS is possible only in separate local volumes.

The melting point of the Ni_3Al phase with a temperature of 1653–1658 K is also included in the pulsation range of 1630–1670 K, corresponding to the temperature profile (Fig. 4, marker of 10 ns), therefore, along with the dominant NiAl phase, a certain amount of the Ni_3Al phase is retained.

Confirmation of the conclusions that the final products of the SHS reaction are dominated by the intermetallic phase NiAl with the bcc-structure and the intermetallic phase Ni_3Al with the fcc-structure is significantly less present, are the data on the numbers of structures of various types at different times, calculated by the Ackland-Jones method using the OVITO program (taking into account the AJ modifier of the OVITO package, [7, 8]) and presented in the table below. The numbers of structures of different types are calculated as follows.

For each atom from the LBS composition, the Ackland-Jones method analyzes the configurations of those neighboring atoms from its "environment", which together with it could form a structure that best suits at least one of the types of the unit crystal cell: fcc, hcp, bcc, ico. The type of the best suitable structure is "assigned" to a given atom, in other words, the counter assigned to this type of structure is increased by one. Therefore, as can be seen from the last column of the table, for each moment in time the total number of structures of various types (the sum for each row of the Table) coincides with the number of all atoms from the LBS composition, equal to 1450915 atoms. The problem of recognizing the corresponding type of unit cell, which includes the analyzed atom with its environment, belongs to the class of "combinatorial" problems.

Let us explain the dynamics of changes over time in the data presented in the Table 1.

Nanosized blocks of Ni and Al atoms in the initial LBS (Fig. 3) are packets of unit crystalline cells of the same fcc-type, which, in the course of the combustion wave propagation, transform into SHS reaction products in the form of intermetallic compounds: $NiAl_3$, Ni_2Al_3, NiAl, Ni_3Al, etc.

Therefore, in the interval of the first 2 ns (see Table 1), a sharp decrease in the number of fcc-type structures is observed, and then the "rate" of a sharp decrease in fcc-type structures slightly slows down. The slowdown in the sharp decrease in fcc-structures occurs due to the fact that, along with the intense growth of the NiAl phase, the formation of the Ni_3Al phase, which also has fcc-type, occurs. However, it should be noted that the formation of the Ni_3Al phase occurs to a lesser extent in comparison with the NiAl phase. As can be seen from the Table 1, starting from 8 ns (possibly starting from 7 ns), the subsequent data practically do not change. In addition, analyzing Figs. 4, 5, 6, 7 and 8, in the time range from 6 to 7 ns, a transition from combustion regime to "aftercombustion (afterburning)" regime occurs. As mentioned above, the presence of hcp-aglomerations of Ni atoms in LBS (more than 22% according to the Table 1) after the first stage end of "combustion and afterburning" and the transition to the second (final) stage of "structure formation" should gradually change the composition and structure of the final SHS products due to secondary physical and chemical transformations.

Table 1. The numbers of unit cells of various types (fcc, hcp, bcc, ico and "other") and their percentages corresponding to the sequence of time moments during the propagation of the combustion wave in the layered-block structure of Ni-Al.

Time, ns	fcc	hcp	bcc	ico	Other	Total sum
0.5	1055753 72.76%	119102 8.21%	97946 6.75%	3512 0.25%	174602 12.03%	1450915 100%
1.0	943842 65.05%	141562 9.76%	117632 8.11%	5975 0.41%	241904 16.67%	1450915 100%
2.0	743416 51.24%	187302 12.91%	156371 10.78%	12843 0.89%	350983 24.18%	1450915 100%
4.0	452641 31.20%	237860 16.39%	215410 14.85%	14950 1.03%	530054 36.53%	1450915 100%
5.0	309692 21.34%	262313 18.08%	253653 17.49%	16672 1.15%	608585 41.94%	1450915 100%
6.0	98659 6.80%	308061 21.23%	298336 20.56%	21198 1.46%	724661 49.95%	1450915 100%
8.0	34045 2.35%	327824 22.59%	283291 19.52%	23116 1.59%	782639 53.94%	1450915 100%
10.0	32561 2.24%	329841 22.73%	282059 19.45%	23205 1.60%	783249 53.98%	1450915 100%
12.0	32562 2.24%	328937 22.67%	281879 19.44%	23128 1.59%	784409 54.06%	1450915 100%
14.0	32576 2.25%	328641 22.65%	282827 19.49%	23146 1.60%	783725 54.01%	1450915 100%
16.0	32606 2.25%	329914 22.74%	282214 19.45%	22780 1.57%	783401 53.99%	1450915 100%
17.0	32441 2.24%	328387 22.63%	282426 19.48%	23196 1.60%	784465 54.05%	1450915 100%

The second stage of SHS - the stage of "structure formation" - in reality, in time instead of tens of nanoseconds (for the nanosized LBS investigated in this work) lasts from several tens of seconds or more. Consequently, in these computational experiments, this stage of SHS was not simulated, since the time that would have to be spent on such a simulation is very long and computational experiments cannot be performed on the computational resources available to the authors of the paper.

However, reasoning theoretically, if we wait for the stage of "structure formation" in the process of cooling of the reacted system, then we can predict a decrease in the proportion of structures corresponding to the following types: hcp, ico and "other" (see Table 1). Consequently, the number of bcc-structures (corresponding to the intermetallic phase NiAl) and fcc-structures (corresponding to the Ni_3Al intermetallic phase) should increase.

The authors of the paper have implemented an additional tool for the "structural" analysis of phase formation in the SHS process by calculating the so-called "substance density profiles" in the reacting medium (in the volume of the LBS). To calculate such density profiles, we used the technique of dividing the LBS volume into small domains, similar to those in calculating temperature profiles (see above). The dimensions of the domains are $4 \times 2.5 \times 16.5$ nm, respectively, along the X, Y, and Z axes. The ratio of the calculated total mass of all atoms contained in the domain to the domain volume of 165 nm^3 estimates the "averaged" value of the substance density in the domain. Thus, let us define the concept of "substance density profile" of LBS (Fig. 9) as a sequence of substance density values corresponding to a sequence of domains along the X direction (to a sequence of X-coordinates of the centers of domains).

With the help of the program developed by the authors, the profiles of the density of matter (Fig. 9), corresponding to the sequence of moments of time, were calculated. At the end of the 1st stage of simulation (after 0.4 ns), when the initial LBS (Fig. 3) has been "relaxed" at a constant temperature of 800 K, the density value averaged over the entire LBS volume turned out to be equal to 6.35 g/cm^3. At the second stage (upon initiation of SHS within 0.1 ns, i.e. by the end of 0.5 ns), within 50 nm of the initial zone of the LBS, a combustion wave arises and a rise in temperature is observed in this zone (Fig. 4). An increase in temperature leads to the formation, first of all, of the NiAl$_3$ phase (in the temperature range from 800 to 1127 K) and then the Ni$_2$Al$_3$ phase (in the temperature range from 800 to 1405–1406 K), followed by the decomposition of the NiAl$_3$ phase at temperatures above 1127 K [12, 13].

The first phase NiAl$_3$ corresponds to an orthorhombic unit cell containing four Ni atoms and twelve Al atoms. In this case, the NiAl$_3$ phase corresponds to the lowest density of 3.9 g/cm^3 [14]. The second phase, Ni$_2$Al$_3$, crystallizes in a trigonal syngony: Al atoms form a pseudocubic structure, Ni atoms occupy 2/3 of the pseudocube centers, and the rest remain vacant. Vacant places lie in planes perpendicular to the trigonal axis. The region of homogeneity of the NiAl$_3$ compound is very narrow, and the width of the region of the Ni$_2$Al$_3$ compound is slightly larger and is within 4 at.% of Ni. The density of the Ni$_2$Al$_3$ compound is about 4.76 g/cm^3 [14]. It should be noted that the densities of the NiAl$_3$ and Ni$_2$Al$_3$ phases formed first of all are significantly lower than the density value of 6.35 g/cm^3, which corresponds to the average value of the LBS density prior to SHS initiation. Therefore, in the initial section of the density profile (Fig. 9, marker of 0.5 ns, within 35 nm along the X axis), a lower density value is observed in the range of 6.08-6.1 g/cm^3 in comparison with the average value of 6.35 g/cm^3, characteristic of the subsequent section of this density profile (from 35 nm to 420 nm along the X axis).

The retention of the average density of 6.35 g/cm^3 in the LBS-volume, starting from the 35 nm mark and up to 420 nm along the X axis, can be explained by the fact that the intense combustion wave has not yet arrived into this LBS-volume, which is confirmed by a small temperature change from 800 to 850 K (Fig. 4, marker of 0.5 ns). In the specified volume of LBS under such conditions, the following stable phases are synthesized in small amounts: NiAl$_3$, Ni$_2$Al$_3$, NiAl (with a density of 5.9-6.02 g/cm^3, [14]) and Ni$_3$Al (with a density of 7.29–7.5 g/cm^3, [14, 15]). Therefore, in the corresponding section of the density profile (Fig. 9, marker of 0.5 ns) pulsations of density values from 6.12–6.15 to 6.4–6.45 g/cm^3 are observed (pulsations of a higher frequency than for density profiles

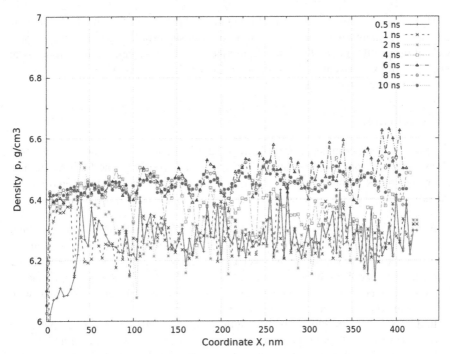

Fig. 9. The density profiles of a substance at successive times during the propagation of a combustion wave in the layered-block structure Ni-Al.

with markers of 6–10 ns). The presence of density peaks exceeding the average value of 6.35 g/cm^3 indicates local formations of NiAl and Ni_3Al phases, and the presence of pulsation minima in the range of 6.12–6.15 g/cm^3 indicates local formations of $NiAl_3$ and Ni_2Al_3 phases. The Ni_3Al phase can be formed in two versions [12, 13]:

$$3Ni + Al \rightarrow Ni_3Al, \quad 2Ni + NiAl \rightarrow Ni_3Al. \tag{3}$$

After 0.5 ns, in the combustion wave zones, the temperature reaches the range of 1630–1700 K; therefore, after the $NiAl_3$ phase, the Ni_2Al_3 phase also decomposes (the $NiAl_3$ phase decomposes above 1127 K, and the Ni_2Al_3 phase – above 1405 K). The decomposition occurs according to the scheme: $2NiAl_3 \rightarrow Ni_2Al_3 + 3Al$ and $Ni_2Al_3 \rightarrow 2NiAl + Al$. Thus, the quantitative growth of the NiAl phase continues with a sharp quantitative decrease in the $NiAl_3$ and Ni_2Al_3 phases, as a result of which the frequency and amplitude of pulsations on subsequent density profiles (with markers of 4, 6, 8, and 10 ns) decrease, and the ranges of the density values shift upward over time. X-coordinates of minima and maxima for density profiles with markers of 6, 8 and 10 ns to a greater extent coincide. The best consistency of extrema is performed for profiles with 8 and 10 ns markers, since, judging by the corresponding thin sections in Figs. 5 and 6 (as opposed to a thin section with marker of 6 ns), the combustion wave passed to the end of the LBS. In other words, as a result of reaching the final mark of 420 nm by the combustion wave, the process of dissolution of components in the Ni-Al system is completed and the layered-block character of the LBS structure is no longer observed.

Thus, analyzing all the density profiles marked with time markers in Fig. 9, it can be seen that with increasing time, passing from one density profile to another, the average levels of the profiles gradually "rise". Consequently, the $NiAl_3$ and Ni_2Al_3 phases with lower densities underwent decomposition, while the NiAl and Ni_3Al phases with higher densities have been remained.

4 Conclusions

Computational experiments using the method of "molecular dynamics simulation (MDS)" showed some features and differences in the course of the SHS reaction in the layered-block atomic structure of Ni-Al (Fig. 3) compared to the layered atomic structure of Ni-Al (Fig. 2). Namely:

1. In the Ni-Al layered-block structure, the velocity of the combustion wave front (and the reaction rate of the SHS) is approximately two times higher than the velocity of the combustion wave front in the layered Ni-Al structure. This effect is explained by the higher specific contact area in LBS between blocks of Ni atoms with blocks of atoms Al in comparison with the specific contact surface of layers of Ni atoms with layers of Al atoms in a layered Ni-Al structure.
2. In the Ni-Al layered-block structure in the process of MD simulation, a "pulsating (oscillatory)" mode of propagation of the SHS combustion wave was found, which is confirmed by the calculated sets of temperature profiles (Fig. 4) and substance densities (Fig. 9). For the "pulsating" mode, the following conditions for SHS initiation were set: stoichiometric ratio $N_{Ni}/N_{Al} = 2.164$, initial heating of the LBS at 800 K, SHS initiation from 800 to 1400 K.
3. The efficiency of the software procedures for calculating temperature profiles and density profiles of a substance, developed by the authors, has been demonstrated, which made it possible to:

   - perform the temperature analysis of SHS microkinetics more accurately;
   - clarify the order of sequence of SHS metallochemical reactions and to recognize the intermetallic phases formed during SHS;
   - study the kinetics of combustion, as well as the evolution of the discrete decay of the thermal structure of the SHS combustion wave in locally unstable microheterogeneous combustion modes.

It should be noted that the number of particles N and the total simulation time are limited by the computing power of the computing system used in the research. The authors of the article used a cluster of 15 PCs with 4-core processors, parallelizing calculations in each computational experiment on 60 parallel threads. Due to the fact that the LAMMPS software package used in the C++ version of the language together with the MPI interface makes it possible to perform parallel computations, computational experiments were carried out to simulate the SHS process in an atomic layered-block structure of Ni-Al with a sufficiently large total number of atoms (almost 1.5 million atoms). In addition, thermal and microstructural analyzes of structural-phase transformations of intermetallic phases during SHS were carried out using the OVITO package.

Under these conditions, the following estimate of the of computer simulation performance of the SHS process was achieved – for a day of continuous computation time on a cluster of 15 PCs, the SHS process was simulated with a duration of almost 2 ns in real time.

Acknowledgments. The work has been supported by Grant № 18-41-220004 provided by the Russian Foundation for Basic Research (RFBR).

References

1. Yvon, P., Carré, F.: Structural materials challenges for advanced reactor systems. J. Nucl. Mater. **385**, 217–222 (2009)
2. Merzhanov, A.G.: Solid-Flame Combustion. Publisher "Izd. ISMAN", Chernogolovka (2000). (in Russian)
3. Verlet, L.: Computer «experiments» on classical fluids. I. Thermodynamical properties of Lennard-Jones molecules. Phys. Rev. **159**, 98 (1967)
4. Verlet, L.: Computer «experiments» on classical fluids. II. Equilibrium correlation functions. Phys. Rev. **165**, 201 (1967)
5. Purja, P.G.P., Mishin, Y.: Development of an interatomic potential for the Ni-Al system. Philos. Mag. **89**(34–36), 3245–3267 (2009)
6. Plimpton, S.: Fast parallel algorithms for short-range molecular dynamics. J Comp. Phys. **117**, 1–19 (1995)
7. Stukowski, A.: Visualization and analysis of atomistic simulation data with OVITO – the open visualization tool. Model. Simul. Mater. Sci. Eng. **18**, 015012 (2010)
8. Ackland, G.J., Jones, A.P.: Applications of local crystal structure measures in experiment and simulation. Phys. Rev. B. **73**(5), 054104 (2006)
9. Shmakov, I.A., Jordan, V.I., Sokolova, I.E.: Computer simulation of SH-synthesis of nickel aluminide by the molecular dynamics method in the LAMMPS package with using of parallel computing. High-Perform. Comput. Syst. and Technol. **2**(1), 48–54 (2018). (in Russian)
10. Mishin, Y., Mehl, M.J., Papaconstantopoulos, D.A.: Embedded-atom potential for B2-NiAl. Phys. Rev. B. **65**(22), 224114 (2002)
11. Turlo, V., Politano, O., Baras, F.: Microstructure evolution and self-propagating reactions in Ni-Al nanofoils: an atomic-scale description. J. Alloys Compd. **708**, 989–998 (2017)
12. Kovalev, O.B., Neronov, V.A.: Metallochemical analysis of the reaction in a mixture of Nickel and Aluminum powders. J. Combust. Explos. Shock. Waves **40**(2), 172–179 (2004)
13. Kovalev, O.B., Belyaev, V.V.: Mathematical modelling of metallochemical reactions in a two-species reacting disperse mixture. J. Combust. Explos. Shock. Waves **49**(5), 563–574 (2013)
14. Nikolsky, B.P., et al. (eds.): Chemist's Handbook, 2nd edn., vol. 1. Publisher "Chemistry", Moscow-Leningrad (1966). (in Russian)
15. Khablov, E.N., Ospennikova, O.G., Bazyleva, O.A.: Cast structural alloys based on nickel aluminide. Motor **4**, 22–26 (2010). (in Russian)

Mathematical Simulation of a Heat Transfer Process in Phase Change Materials

Sergey Markov[1]([✉]) [iD], Ella Shurina[2] [iD], and Natalya Itkina[3] [iD]

[1] Trofimuk Institute of Petroleum Geology and Geophysics of SB RAS, Koptug Avenue 3, 630090 Novosibirsk, Russia
www.sim91@list.ru
[2] Novosibirsk State Technical University, Karla Marx Avenue 20, 630073 Novosibirsk, Russia
[3] Institute of Computational Technologies of SB RAS, Academician M.A. Lavrentiev Avenue 6, 630090 Novosibirsk, Russia

Abstract. In the paper, we present a computational scheme for mathematical simulation of heat transfer processes in phase-changing multiscale media. In this problem, the correct approximation of heat flux jumps on phase transition boundaries is the main difficulty. Our approach is based on using a nonconforming multiscale finite element method to solve Stefan's problem. We propose to divide a solution of Stefan's problem into two components. A discontinuous component is determined in phase transition zones (fine level). The discontinuous component is approximated by a discontinuous Galerkin method. A continuous component is determined everywhere (coarse level). The continuous component is approximated by the classic finite element method. In this approach, a discrete analogue of Stefan's problem can be solved in parallel. For the correct approximation of the heat flux jump on the phase transition boundary, we introduce a special lifting operator in the variational formulation of the multiscale discontinuous Galerkin method. Results of verification procedure for the developed computational scheme are shown using Stefan's problem with the analytical solution. A validation procedure is performed using a comparative analysis of mathematical simulation results with physical experimental data. In the physical experiment, a phase change material sample was heated. At discrete time moments, the temperature was recorded using a sensor. We performed the mathematical simulation of the heat transfer process in the phase change material sample using experiment conditions. The difference between the calculated temperature field and physical experiment data was less than 5%.

Keywords: Phase change · Stefan's problem · Discontinuous Galerkin method

1 Introduction

One of the problems facing modern society is the rational use of natural and energy resources today. For solving the problem, it is necessary to develop new phase change materials (PCM) and to study heat transfer processes in media with phase transitions [1].

© Springer Nature Switzerland AG 2020
V. Jordan et al. (Eds.): HPCST 2020, CCIS 1304, pp. 62–79, 2020.
https://doi.org/10.1007/978-3-030-66895-2_5

The composite PCM are characterized by a multiscale geometric structure and high contrast of physical properties [2]. Research methods should take into account the features of such systems. Mathematical simulation is one of the suitable methods to study the heat transfer process in the PCM.

Mathematical models of heat transfer processes with phase transitions are based on Stefan's problem [3]. There are many computational schemes for solving the Stefan problem.

Meshfree finite element methods, Boltzmann's lattice method, and natural neighbor methods are meshfree methods [4–7]. In the meshfree methods, a medium is considered at the level of a single particle. To implement the meshless methods, it is necessary to develop effective parallel algorithms using powerful multiprocessor systems.

There are various computational schemes of finite difference methods (FDM), finite volume methods (FVM), and finite element methods (FEM). In this mesh-based methods, special technologies are realized to solve problems with phase transitions. For example, tracking front methods (methods of fixed points [8], approaches using the form derivatives to minimize error functionals [9]), implicitly accounting front methods (method of setting the level [10], enthalpy method [11], phase field method [12]) are the most popular technologies for taking into account the phase transition.

For heat transfer problems, finite volume methods are universal tools for constructing conservative computational schemes. The use of special schemes for approximating conservation laws in an integral form makes it possible to obtain a stable solution to a wide class of problems [13].

Discontinuous Galerkin methods (DG) are members of nonconforming finite element methods. The discontinuous Galerkin method is one of the suitable methods for solving a wide class of singular problems. The DG method combines the efficiency of finite volume methods with computational flexibility of finite element methods. Applications of the DG method for solving Stefan's problem can be found in [14, 15].

For mathematical simulating the heat transfer processes in multiscale media, computational schemes based on a multiscale finite element method (MsFEM) have proven themselves well. In the MsFEM, multiscale problems are divided into many smaller ones, which are solved in parallel [16].

Virtual finite element methods (VFEM) are members of the multiscale finite element methods. The VFEM uses polyhedral meshes with nonpolynomial shape functions [17].

To solve problems with moving boundaries, an extended finite element method (XFEM) is used. Today, XFEM technologies are designed for a limited set of applications [18, 19].

Our approach applies the idea described in [16] to discretize the stationary diffusion equation. In [20], we extended this concept to design and implement an algorithm for solving Stefan's problem. In this paper, for the correct approximation of the heat flux jump on the phase transition boundary, we introduce a special lifting operator in the variational formulation of the multiscale discontinuous Galerkin method from. The main importance of the article lies in the validation procedure of the developed computational scheme.

2 Problem Definition

We consider a parallelepiped sample with a height $L_x = 0.02$ m, length $L_y = 0.1$ m and width $L_z = 0.1$ m. The sample consists of inclusions (diatomite fine fraction) impregnated with paraffin. The paraffin phase transition temperature is 47 °C. A matrix of the sample is polyurethane.

The sample physical properties are given in Table 1.

Table 1. Physical properties of materials.

Material		Thermal conductivity λ, W/(m•K)	Density ρ, kg/m^3	Specific heat c, J/(kg•K)	Specific heat of phase transition, L, kJ/kg
Polyurethane		0.315	1210	1470	–
Diatomite		0.131	683	840	–
Paraffin	Solid-phase	0.3	800	2190	160
	Liquid-phase	0.27		2130	

Figures 1 and 2 show the sample structure with inclusions impregnated with paraffin and with inclusions that are not impregnated with paraffin. The edging of diatomite fine fraction impregnated with paraffin is presented in Fig. 3.

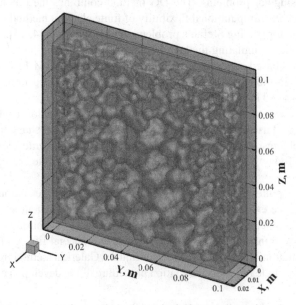

Fig. 1. Sample structure: matrix – polyurethane, inclusions – diatomite edged with a paraffin saturation zone.

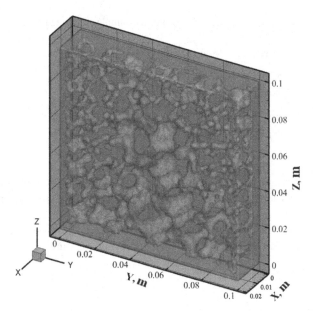

Fig. 2. Sample structure: matrix – polyurethane, inclusions – diatomite without a paraffin saturation zone.

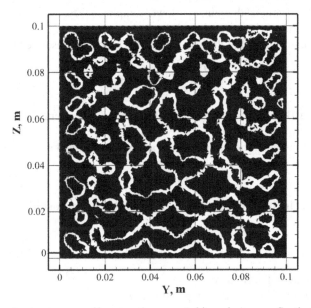

Fig. 3. Sample structure: paraffin saturation zone (white color), cross Section X = 0.01.

The volumetric content of the phase change material sample components is shown in Table 2.

Table 2. Volumetric content of the sample components.

Material	Volume, m^3	Volume concentration, %
Polyurethane	0.0001382	69.1
Diatomite	0.0000272	13.6
Diatomite	0.0000346	17.3
Sample	0.0002	–

The initial temperature of the sample and environment is T_0. On the lower sample surface, the boundary temperature is T_1. The side and top sample surfaces are thermally insulated.

The computational problem is a mathematical simulation of nonstationary temperature fields over the sample volume at given T_0 and T_1 temperature values.

3   Methods

3.1 Mathematical Model

In this section, we formulate Stefan's problem.

Let Ω be computational domain, S_1 – lower sample surface, S_2 – thermally insulated sample surfaces, $\xi(t)$ – phase transition boundary as a function of time.

The mathematical model of the heat conduction process is described by the equation

$$\rho c \frac{\partial T}{\partial t} = \nabla \cdot (\lambda \nabla T) \text{ in } \Omega, \tag{1}$$

where ρ – density [kg/m^3], c – specific heat [J/kg • K], T – temperature [K], λ – thermal conductivity [W/(m • K)].

At the initial moment, the sample temperature is determined as

$$T|_{t=0} = T_0. \tag{2}$$

On the thermally insulated sample surfaces, the boundary condition for the Eq. (1) is formulated as

$$\lambda \nabla T \cdot \mathbf{n}|_{S_2} = 0, \tag{3}$$

where $\mathbf{n}$ – unit external normal vector.

On the lower sample face, the temperature is T_1

$$T|_{S_1} = T_1. \tag{4}$$

On the phase transition boundary $\xi(t)$, the ideal contact condition is defined as

$$[T]|_{\xi(t)} = 0, \tag{5}$$

and heterogeneous conditions for the heat flux jump are formulated as

$$[\lambda \nabla T]|_{\xi(t)} = -\rho L \frac{\partial \xi(t)}{\partial t}, \tag{6}$$

where L – specific heat of phase transition [J/kg]. The combination of the steps required to solve Stefan's problem listed in the Table 3.

Table 3. Algorithm for solving Stefan's problem

1	To solve the problem (1)–(4) for the initial position of the phase transition front and with uniform conjugation conditions (6)
2	To determine a new position of the phase transition front
3	To solve the problem (1)–(4) for the next layer in time, taking into account the new front position
4	To repeat step 2 until the change in the front position is greater than a certain constant ε

3.2 Notations

Let $\Omega \subset R^3$ be a computational domain, $\Xi_h(\Omega)$ be a union of cells Ω_k, $\Gamma = \bigcup\limits_k \partial \Omega_k$ be a set of cells borders, $\Gamma_0 = \Gamma \backslash \partial \Omega$ be a set of inner cells borders, and $T(\Gamma) = \prod\limits_{\Omega_k \in \Xi_h(\Omega)} L_2(\partial \Omega_k)$ is a space of function traces. On the set $\Xi_h(\Omega)$, the finite-dimensional function subspaces are defined as [16]

$$V^h = \{v | v \in L_2(\Omega) : v \in P_m(\Omega_k)\}, \; V^h \subset L_2(\Omega), \tag{7}$$

$$V_1^h = \left\{ v | v \in H^1(\Omega) : v \in P_m(\Omega_k) \right\}, \; V_1^h \subset H^1(\Omega), \tag{8}$$

$$\mathbf{V}^h = \left\{ \mathbf{v} | \mathbf{v} \in [L_2(\Omega)]^3 : \mathbf{v} \in [P_m(\Omega_k)]^3 \right\}, \; \mathbf{V}^h \subset [L_2(\Omega)]^3, \tag{9}$$

where $P_m(\Omega_k)$ – the space of polynomials with degree m defined in the cell Ω_k.

To construct trace operators on the inter-element boundary, the average $\{\cdot\}$ and jump $[\cdot]$ are introduced. For the functions $\mathbf{v} \in [T(\Gamma)]^3$ and $v \in T(\Gamma)$ on the external boundary $\partial \Omega$, we can write [14]

$$[\cdot] : [T(\Gamma)]^3 \to L_2(\Gamma), \; [\mathbf{v}]|_{\partial \Omega} = \mathbf{v} \cdot \mathbf{n},$$
$$\{\cdot\} : [T(\Gamma)]^3 \to [L_2(\Gamma)]^3, \; \{\mathbf{v}\}|_{\partial \Omega} = \mathbf{v},$$
$$[\cdot] : T(\Gamma) \to [L_2(\Gamma)]^3, \; [v]|_{\partial \Omega} = v\mathbf{n},$$
$$T(\Gamma) \to L_2(\Gamma), \; \{v\}|_{\partial \Omega} = v, \tag{10}$$

on the inner boundary $\Gamma_0 = \partial \Omega_k \cap \partial \Omega_n$, we have [14]

$$[\cdot] : [T(\Gamma)]^3 \to L_2(\Gamma_0), \; [\mathbf{v}]|_{\Gamma_0} = \mathbf{v}_k \cdot \mathbf{n}_k + \mathbf{v}_n \cdot \mathbf{n}_n,$$

$$\{\cdot\} : [\mathrm{T}(\Gamma)]^3 \to [L_2(\Gamma_0)]^3, \ \{\mathbf{v}\}\big|_{\Gamma_0} = (\mathbf{v}_k + \mathbf{v}_n)/2,$$

$$[\cdot] : \mathrm{T}(\Gamma) \to [L_2(\Gamma_0)]^3, \ [v]\big|_{\Gamma_0} = v_k\mathbf{n}_k + v_n\mathbf{n}_n,$$

$$\mathrm{T}(\Gamma) \to L_2(\Gamma_0), \ \{v\}\big|_{\Gamma_0} = (v_k + v_n)/2, \tag{11}$$

where the lower index indicates belonging to Ω_i.

For the correct approximation of the heat flow jump, we introduce the lifting operator

$$r : \mathrm{T}(\Gamma) \to [L_2(\xi(t))]^3, \ \int_{\Omega_m} r(v)\psi d\Omega_m = -\int_{\xi(t)} v[\psi] d\xi(t), \ \forall \psi \in \mathbf{V}^h, \tag{12}$$

where Ω_m – subdomain containing the phase transition front $\xi(t)$.

3.3 Multiscale Discontinuous Galerkin Method for Stefan's Problem

Let $\Omega = \Omega^C \cup \Omega^D$ and $T^h = T^h_C + T^h_D$, where T^h_C is the continuous solution component of Stefan's problem determined over in Ω (coarse level), T^h_D is the discontinuous solution component determined in Ω^D only (fine level). On the boundary $\partial\Omega^D$, the discontinuous component vanishes

$$T^h_D\big|_{\partial\Omega^D} = 0. \tag{13}$$

The variational formulation of the multiscale discontinuous Galerkin method for Stefan's problem (1)–(6) is defined as: find $T^h_C \in V^h_1 \times [0, \mathrm{T}]$ and $T^h_D \in V^h \times [0, \mathrm{T}]$ that $\forall \psi^h_C \in V^h_1$ and $\psi^h_D \in V^h$

$$a\left(T^h_C, \psi^h_C\right) + a\left(T^h_D, \psi^h_C\right) = \left(f^h, \psi^h_C\right),$$

$$a\left(T^h_C, \psi^h_D\right) + a\left(T^h_D, \psi^h_D\right) = \left(f^h, \psi^h_D\right), \tag{14}$$

where bilinear and linear forms are defined as

$$a\left(T^h, \psi^h\right) = \sum_{\Omega_k} \left(\int_{\Omega_k} c\rho\frac{\partial T^h}{\partial t}\psi^h d\Omega + \int_{\Omega_k} \lambda\nabla T^h \cdot \nabla\psi^h d\Omega_k \right)$$

$$+ \sum_{\partial\Omega_k\backslash\xi(t)\,\partial\Omega_k} \int \lambda([T^h] \cdot \{\nabla\psi^h\} - \{\nabla T^h\} \cdot [\psi^h]) \, dS + \sum_{\Omega_m} \left(\int_{\Omega_m} r([\lambda\nabla T^h]) \cdot \nabla\psi^h d\Omega_k \right), \tag{15}$$

$$\left(f^h, \psi^h\right) = \int_{S_1} \left(\lambda T_1 \nabla\psi^h \cdot \mathbf{n}\right) dS, \tag{16}$$

where T_1 – temperature on the surface S_1.

To take into account the Stefan interface condition (6), we modify the integrals in the bilinear form (15) as

$$\int\limits_{\xi(t)} \lambda \{\nabla T^h\} \cdot [\psi^h] dS = \int\limits_{\xi(t)} \frac{\lambda^s \nabla T^s - \lambda^l \nabla T^l}{2} \psi^h \cdot \mathbf{n} d\xi(t) = -\int\limits_{\xi(t)} \frac{1}{2} \rho V L \psi^h d\xi(t),$$

(17)

$$\sum_{\Omega_m} \left(\int\limits_{\Omega_m} r([\lambda \nabla T^h]) \cdot \nabla \psi^h d\Omega_k \right) = -\int\limits_{\xi(t)} [\lambda \nabla T^h][\nabla \psi^h] d\xi(t)$$
$$= \int\limits_{\xi(t)} \rho V L (\nabla \psi^h \cdot \mathbf{n}) d\xi(t),$$

(18)

where V – modulus of the front velocity vector.

The linear form (16) is modified as

$$\left(f^h, \psi^h\right) = \int\limits_{S_1} \lambda T_1 \nabla \psi^h \cdot \mathbf{n} dS + \int\limits_{\xi(t)} \frac{1}{2} \rho V L \psi^h d\xi(t) + \int\limits_{\xi(t)} \rho V L \left(\nabla \psi^h \cdot \mathbf{n}\right) d\xi(t). \quad (19)$$

The new position of the phase transition boundary is defined in the form

$$\xi(t + \Delta t) = \xi(t) - \frac{\Delta t}{\rho L} [\lambda \nabla T]|_{\xi(t+\Delta t)}. \quad (20)$$

3.4 Discrete Analogue of Stefan's Problem

Let $T_C^h \in V_1^h$ and $T_D^h \in V^h$ be the finite element interpolants

$$T_D^h = \sum a_i \psi_i^D, \ \psi_i^D \in P^h \text{ in } \Omega^D,$$
$$T_C^h = \sum b_i \psi_i^C, \ \psi_i^C \in P^h \text{ in } \Omega, \quad (21)$$

where $\{\psi_i^C\}|_\Gamma = \psi_i^C, \ [\psi_i^C]|_\Gamma = 0.$

To approximate the time derivative, we use the finite difference technology

$$\frac{\partial T^h}{\partial t} = \frac{T^h(t + \Delta t) - T^h(t)}{\Delta t}, \quad (22)$$

where Δt – time step.

The bilinear form (15) and linear form (16) define in the discrete case a matrix and vector whose elements are calculated by the formulas

$$[\mathbf{A}]_{ij}^{\Omega_K} = \int\limits_{\Omega_k} c\rho \frac{\psi_j^h}{\Delta t} \psi_i^h d\Omega + \int\limits_{\Omega_k} \lambda \nabla \psi_j^h \cdot \nabla \psi_i^h d\Omega_k$$
$$+ \int\limits_{\partial\Omega_k} \lambda \left([\psi_j^h] \cdot \{\nabla \psi_i^h\} - \{\nabla \psi_j^h\} \cdot [\psi_i^h]\right) dS - \int\limits_{\xi(t)} [\lambda \nabla \psi_j^h][\nabla \psi_i^h] d\xi(t),$$

(23)

$$[\mathbf{F}]_i^{\Omega_K} = \int\limits_{\Omega} c\rho \frac{T^h(t)}{\Delta t} \psi_i^h d\Omega + \int\limits_{S_1} \lambda T_1 \nabla \psi_i^h \cdot \mathbf{n} dS$$

$$+ \int\limits_{\xi(t)} \frac{1}{2}\rho VL\psi_i^h d\xi(t) + \int\limits_{\xi(t)} \rho VL\nabla\psi_i^h \cdot \mathbf{n} d\xi(t). \tag{24}$$

To approximate the temperature field, a second-order hierarchical basis of the H^1 space on the tetrahedra is used. An algorithm for constructing the hierarchical basis of the H^1 space can be found in [21].

The discrete analog of Stefan's problem (14) can be solved in parallel using domain decomposition methods or algebraic multigrid solvers.

4 Results

In the section, we present results of verification and validation procedures for the computational scheme.

4.1 Verification Procedure of the Computational Scheme

Let $\Omega_l = [0, 1] \times [0, 1] \times [0, 1]$ and $\Omega_S = [0, 1] \times [0, 1] \times [1, 2]$ be subdomains filled with liquid and solid substance phases, respectively. Figure 4 shows this computational domain.

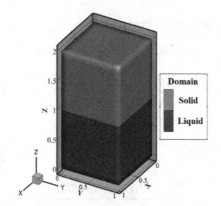

Fig. 4. Computational domain structure in the verification procedure.

At the initial moment, the liquid phase temperature is $T_l(z, 0) = 1$, and the solid phase temperature is $T_s(z, 0) = 0$. On the lower boundary $z = 0$, the temperature is $T_l(0, t) = 1$. Other faces are thermally insulated.

Let K_l and K_s denote the thermal diffusivity coefficients

$$K_l = \frac{\lambda_l}{\rho_l c_l}, \quad K_s = \frac{\lambda_s}{\rho_s c_s}, \tag{25}$$

where λ – thermal conductivity, ρ – density, c – specific heat, K – thermal diffusivity coefficient [m²/s].

The Eq. (1) in the subdomains Ω_l and Ω_s will look like [3]

$$\frac{\partial T_l}{\partial t} = \nabla \cdot K_l \nabla T_l \text{ in } \Omega_l,$$

$$\frac{\partial T_s}{\partial t} = \nabla \cdot K_s \nabla T_s \text{ in } \Omega_s. \tag{26}$$

The initial conditions are formulated as

$$T_l(z, 0) = 1, \ T_s(z, 0) = 0. \tag{27}$$

On the lower boundary $z = 0$, the temperature is determined as

$$T_l(0, t) = 1. \tag{28}$$

Other faces are thermally insulated. Neumann's condition is formulated as

$$K_i \nabla T_i \cdot \mathbf{n}|_{S_2} = 0, \tag{29}$$

where S_2 – thermally insulated faces.

To formulate Stefan's problem, it is necessary to redefine the problem (26)–(29) by the Stefan's conditions

$$T_l|_{\xi(t)} = T_s|_{\xi(t)} = T_M, \tag{30}$$

$$[\lambda \nabla T]|_{\xi(t)} = -\rho L \frac{\partial \xi(t)}{\partial t}, \tag{31}$$

where T_M – phase transition temperature.

Stefan's problem (26)–(31) has an analytical solution. The phase transition front position is determined as [3]

$$\xi(t) = \xi(t_0) + 2\alpha \sqrt{K_l t}, \tag{32}$$

and the temperature distribution is given by the function [3]

$$T(t, z) = \begin{cases} T_l(z, 0), \ z < \xi(t_0), \\ T_l(z, 0) - (T_l(z, 0) - T_M) \dfrac{\text{erf}\left(\frac{z - \xi(t_0)}{2\sqrt{K_l t}}\right)}{\text{erf}(\alpha)}, \ \xi(t_0) \le z \le \xi(t), \\ T_M, z > \xi(t), \end{cases} \tag{33}$$

where $\text{erf}(x) = \frac{2}{\sqrt{\pi}} \int\limits_0^x \exp\left(-s^2\right) ds$, α – solution of the equation [3]

$$\alpha \exp(\alpha^2) \, \text{erf}(\alpha) = \frac{c_l(T_l(z, 0) - T_M)}{L\sqrt{\pi}}. \tag{34}$$

The phase transition front does not reach the upper boundary of the solid phase.

To verify the computational scheme, let us consider the problem of ice melting. We choose the initial position of the phase transition in the plane $z = 1$. Physical parameters of ice and water are given in Table 4.

The ice phase transition temperature is $0\,°C$. Using the formula (34), we have $\alpha = 7.8243e - 02$.

Figure 5 shows the triangulation of computational domain. The triangulation adapts to the phase transition zone.

To approximate the temperature field, the second-order hierarchical basis of the H^1 space is used. The thermal conductivity process is simulated for 10 h with a time step of 1 h. The SLAE solver is based on the GMRES and BiCGStab algorithms with AMG-preconditioner. The exit from the iterative process is carried out according to the residual norm $\varepsilon = 10^{-10}$.

Figure 6 shows the graphs of phase transition front position after solving the problem (26)–(31). Figure 7 shows the graphs of the temperature field distribution after solving the problem (26)–(31) at the times $t = 3600, 18000, 36000$ s.

Table 4. Physical parameters of water and ice

Material	Thermal conductivity λ, W/(m • K)	Density ρ, kg/m^3	Specific heat c, J/(kg • K)	Thermal diffusivity coefficient K, [m^2/s]	Specific heat of phase transition, L, kJ/kg
Ice	2.330	900	2110	$1.2270e - 6$	340
Water	0.556	1000	4180	$1.3301e - 7$	–

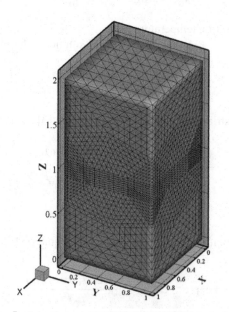

Fig. 5. Adaptive mesh tessellation: 80384 tetrahedra.

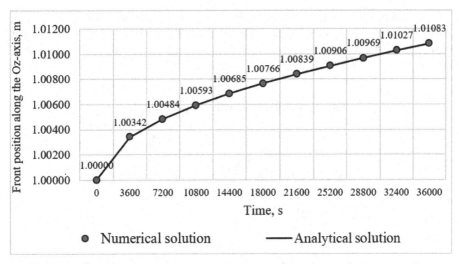

Fig. 6. Changing the phase transition front position over time.

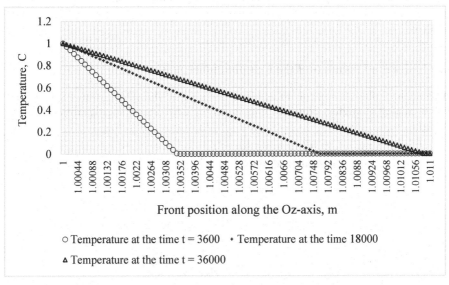

Fig. 7. Temperature change in the phase transition zone over time.

The kinks in the temperature graphs correspond to positions of the phase transition front. For example, a break in the isotherm at the time $t = 3600$ occurs at level $z \approx 1.0034$. At the time $t = 3600$, the position of the phase transition front corresponds to level $z = 1.00342$.

Table 5 shows the $L_2(\Omega)$-norm of relative error in the calculated temperature field. Let $T^{analytical}$ be a temperature calculated by the formula (33), T^h is a numerical solution obtained using the mesh showed in Fig. 5, $T^{h/2}$ is a numerical solution obtained using the third-order hierarchical basis of the H^1 space.

Table 5. Relative error in the temperature field.

Time, s	$\dfrac{\left\| T^{analytical} - T^h \right\|_{L_2(\Omega)}}{\left\| T^{analytical} \right\|_{L_2(\Omega)}}$	$\dfrac{\left\| T^{analytical} - T^{h/2} \right\|_{L_2(\Omega)}}{\left\| T^{analytical} \right\|_{L_2(\Omega)}}$
3600	$3.25e - 10$	$2.87e - 10$
18000	$5.98e - 09$	$9.93e - 10$
36000	$3.18e - 08$	$5.64e - 09$

The computational scheme shows high accuracy in solving Stefan's problem, which has an analytical solution. A decrease in the relative error of numerical solution is observed with an increase in the order of basis functions. Thus, the developed computational scheme guarantees a physically relevant result when solving the formulated problem.

4.2 Validation Procedure of the Computational Scheme

The sample of phase change material (see Fig. 1) is placed in a heat-insulated container with a heating aluminum plate at the bottom. Figure 8 shows the experimental setup structure.

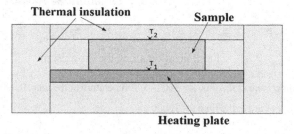

Fig. 8. Experimental setup

The aluminum plate heats the sample in two temperature modes. Figures 9 and 10 show the graphs of temperature modes. The temperature on the lower and upper surfaces of the sample is measured using sensors T_1 and T_2. The maximum temperature in the first heating mode is 65.2 °C. The maximum temperature in the second heating mode is 63 °C. The initial sample temperature is 32 °C.

Physical parameters of the sample are presented in Table 1.

In more detail, the experiment is described in [22].

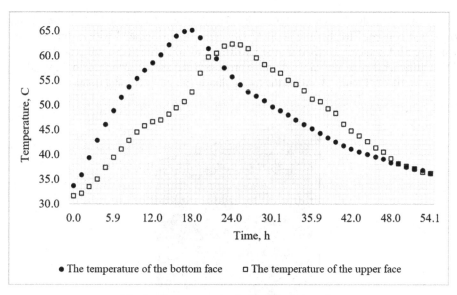

Fig. 9. Temperature in the first heating mode.

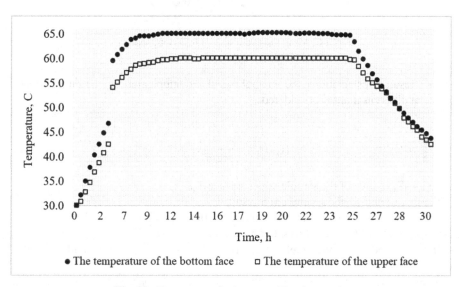

Fig. 10. Temperature in the second heating mode.

4.3 Mathematical Simulation Results

The mathematical model of heat conduction process with a phase transition is described by the problem (1)–(6) at $T_0 = 32\,°C$ and T_1 from the conditions of the physical experiment (See Fig. 9 and 10). For approximating the temperature T_1, cubic interpolation splines are used.

To approximate the temperature field, the second-order hierarchical basis of the H^1 space is used. The thermal conductivity process is simulated with a time step of 1 h. The SLAE solver is based on the GMRES and BiCGStab algorithms with AMG-preconditioner. The exit from the iterative process is carried out according to the residual norm $\varepsilon = 10^{-10}$. Figure 11 shows the triangulation of computational domain.

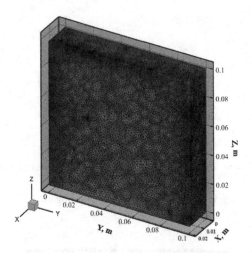

Fig. 11. Mesh tessellation: 2 033 376 tetrahedra.

As a result of calculations, the integral calculated temperature on the sample upper surface at different times is considered

$$T_{int}(t) = \frac{\int\limits_{S_{up}} T(t)dS}{mesS_{up}}, \tag{35}$$

where $mesS_{up}$ – area of the sample upper surface S_{up}.

Figures 12 and 13 show simulation results.

In the first heating mode, a bend in the temperature graph is observed 10 h after the start of heating and 6 h after the end of heating.

In the second mode, the graph is bent 10 h after the start of heating and 12 h after the end of heating.

These bends characterize the beginning and end of the phase transition process in the sample. The obtained simulation results are confirmed with the physical experiment data. It means that the test phase-change material has high accumulating properties.

To estimate the relative error of solution $T_{int}(t)$, we use the vector norm $\|\mathbf{x}\|_\infty = \max\limits_i |x_i|$ in the first heating mode as

$$\delta_1 T_{int}(t) = \frac{\|T_{int}(t) - T_E(t)\|_\infty}{\|T_{int}(t)\|_\infty} 100\% = 4.2, \tag{36}$$

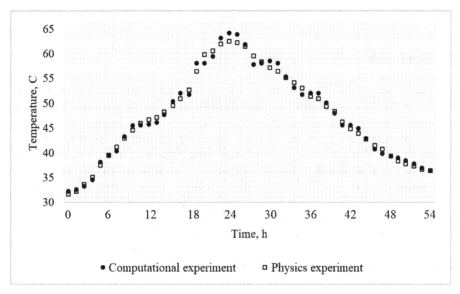

Fig. 12. First heating mode: calculated and practical temperatures on the upper sample surface.

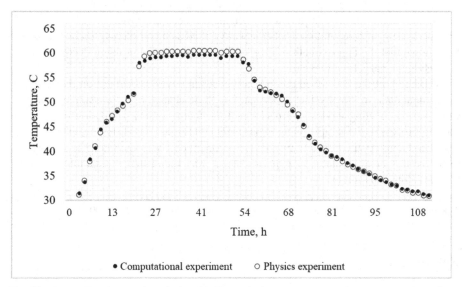

Fig. 13. Second heating mode: calculated and practical temperatures on the upper sample surface

and in the second heating mode as

$$\delta_2 T_{\text{int}}(t) = \frac{\|T_{\text{int}}(t) - T_E(t)\|_\infty}{\|T_{\text{int}}(t)\|_\infty} 100\% = 2.9, \qquad (37)$$

where $T_E(t)$ – temperature in the physics experiment.

We can see the relative error in the simulation results was less than 5%.

5 Conclusion

In this paper, we propose a computational scheme based on the multiscale discontinuous Galerkin method for the mathematical simulation of heat transfer processes with phase transitions.

For solving Stefan's problem, we use a multiscale approach based on splitting the solution into two components. A discontinuous component is determined in phase transition zones (fine level/scale). The discontinuous component is approximated by the discontinuous Galerkin method. A continuous component is determined everywhere (coarse level/scale). The continuous component is approximated by the classic finite element method. In this approach, a discrete analogue of Stefan's problem can be solved in parallel using domain decomposition methods and algebraic multigrid solvers.

An explosive increase in degrees-of-freedoms is a weak point of all computational schemes based on the discontinuous Galerkin method. The multiscale approach makes it possible to approximate Stefan's problem and reduces the number of the degrees-of-freedoms.

To correctly approximate Stefan's condition, a special lifting operator was introduced into the variational formulation of the multiscale discontinuous Galerkin method. As a result, the stability of the computational scheme has been improved.

The developed computational scheme was verified using problems close to real ones. It is shown that the approximation error of the solution to the problem decreases with increasing the basis functions order.

Validation of the developed computational scheme was carried out on the physical experiment data. In the physical experiment, a phase change material sample is heated using various heating modes. We performed the mathematical simulation of the heat transfer process in the sample using experiment conditions. The difference between the calculated temperature field and physical experiment data was less than 5%. According to the results of computational experiments, it was found that the sample has high accumulating capabilities.

Acknowledgments. The research was supported by RSF (project No. 20-71-00134).

References

1. Fang, Y., Nin, J., Deng, S.: Numerical analysis for maximizing effective energy storage capacity of thermal energy storage system by enhancing heat transfer in PCM. Energy Build. **160**, 10–18 (2018)
2. Zeinelabdein, R., Omer, S., Gan, G.: Critical review of latent heat storage system for free cooling in buildings. Renew. Sustain. Energy Rev. **82**, 2843–2868 (2018)
3. Gupta, S.C.: The Classical Stefan Problem: Basic Concepts, Modelling and Analysis, 2nd edn. Gardners Books, Amsterdam (2003)
4. Reutskiy, S.Y.: A meshless method for one-dimensional Stefan problems. Appl. Math. Comp. **217**(23), 9689–9701 (2011)

5. Johansson, B.T., Lesnic, D., Reeve, Th: A meshless regularization method for a two-dimensional two-phase linear inverse stefan problem. Adv. Appl. Math. Mech. **5**(6), 825–845 (2013)

6. Karami, A., Abbasbandy, S., Shivanian, E.: Meshless local Petrov-Galerkin formulation of inverse Stefan problem via moving least squares approximation. Math. Comput. Appl. **24**(4), 101 (2019)

7. Wen-Shu, J., Jeng-Rong, H., Chun-Pao, K.: Lattice Boltzmann method for the heat conduction problem with phase change. Numer. Heat Transf. Part B Fundam. **39**(2), 167–187 (2010)

8. Ramirez, J.C., Beckermann, C.: Examination of binary alloy free dendritic growth theories with a phase-field model. Acta Mater. **53**(6), 1721–1736 (2005)

9. Javierre, E.: A comparison of numerical models for one-dimensional Stefan problems. J. Comput. Appl. Math. **192**(2), 445–459 (2006)

10. Javierre-Perez, E.: Literature study: numerical methods for solving Stefan problems. Reports of the Department of Applied Mathematical Analysis, pp. 1–85(2003)

11. Date, A.W.: A novel enthalpy formulation for multidimensional solidification and melting of a pure substance. Sadhana **19**(5), 833–850 (1994)

12. Singer-Loginova, I., Singer, H.M.: The phase field technique for Modelling multiphase materials. Rep. Prog. Phys. **71**, 106501 (2008)

13. Popov, N., Tabakova, S., Feuillebois, F.: Numerical modelling of the one-phase Stefan problem by finite volume method. In: Li, Z., Vulkov, L., Waśniewski, J. (eds.) NAA 2004. LNCS, vol. 3401, pp. 456–462. Springer, Heidelberg (2005). https://doi.org/10.1007/978-3-540-31852-1_55

14. Pei, Ch., Sussman, M., Hussaini, M.: A space-time discontinuous Galerkin spectral element method for the Stefan problem. Discrete Continuous Dyn. Syst. **B23**(9), 3595–3622 (2018)

15. McLean, W.: Implementation of high-order, discontinuous Galerkin time stepping for fractional diffusion problems. https://arxiv.org/pdf/2003.09805.pdf. Accessed 09 Jan 2020

16. Bochev, P., Hughes, T., Scovazzi, G.: A multiscale discontinuous Galerkin method. Comput. Meth. Appl. Mech. Eng. **195**, 2761–2787 (2006)

17. Brezzi, F., Marini, L.: Virtual element method for plate bending problems. Comput. Meth. Appl. Mech. Eng. **253**, 455–462 (2012)

18. Martina, D., Chaoukia, H., Robert, J.-L., Ziegler, D., Fafarda, M.: A XFEM Lagrange multiplier technique for Stefan problems. Front. Heat Mass Transf. **7**(31), 1–9 (2016)

19. Li, M., Chaouki, H., Robert, J.-L., Ziegler, D., Martin, D., Fafard, M.: Numerical simulation of Stefan problem with ensuing melt flow through XFEM/level set method. Finite Elem. Anal. Des. **148**, 13–26 (2018)

20. Markov, S., Shurina, E., Itkina, N.: A multi-scale discontinuous Galerkin method for mathematical modeling of heat conduction processes with phase transitions in heterogeneous media. J. Phys: Conf. Ser. **1333**, 032052 (2019)

21. Solin, P., Segeth, K., Dolezel, I.: Higher-Order Finite Element Methods, 1st edn. Chapman and Hall/CRC, New York (2003)

22. Nizovtsev, M.I., Borodulin, V.Y., Letushko, V.N., Terekhov, V.I., Poluboyarov, V.A., Berdnikova, L.K.: Thermophys. Aeromech. **26**(3), 313–324 (2019)

Modeling of Polycrystalline Materials Deformation with Dislocation Structure Evolution and Transition to Fracture

Kseniia Kurmoiartseva$^{(\boxtimes)}$ ⓘ, Natalya Kotelnikova ⓘ, and Peter Trusov ⓘ

Perm National Research Polytechnic University, Komsomolskiy Ave. 29, 614990 Perm, Russia
kurmoiartseva.k@mail.ru, kotelnickova@gmail.com,
tpv@matmod.pstu.ac.ru

Abstract. Computing experiments on simulation of deformation and destruction processes of polycrystalline materials is a relevant and efficient research method. This approach suggests building a mathematical model of the material and its exploitation for numerical experiments. The development of the physically based mathematical model for the description of the material behavior during deformation will allow optimizing the properties of constructions without conducting numerous natural experiments. It has been established that when solving similar tasks, it is necessary to take into account the evolution of the meso- and microstructure, including nucleation and development of defects at various structural and scale levels. Consideration of the dislocation structure evolution enables establishing the regions of dislocation accumulation and, as a consequence, simulating the nucleation and subsequent development of microcracks. The purpose of the present study is to develop and realize the dislocation-oriented direct elasto-visco-plastic model, taking into account the processes of microcrack nucleation. The structure and the algorithm of the numerical implementation of the direct model have been presented; the structural levels of the description of elasto-plastic deformation have been identified; a system of evolution equations for the description of dislocation motions and dislocation substructures formation on slip systems with subsequent transition to destruction has been presented. Using the parameters, characterising the dislocation structure, a criterion of the transition into the destructed state has been suggested, in which the material element of the appropriate scale level loses an ability to resist external impacts. The description of the application software package intended for the implementation of multilevel models of the representative volume of polycrystalline solids has been given. The submodel of the description of the dislocation density evolution of the crystallite has been realised; the influence of the mechanisms of nucleation and annihilation of dislocations on their total density has been analysed.

Keywords: Numerical simulation · Algorithms · Dislocation-Oriented model · Crystal plasticity · Dislocation densities · Microcrack nucleation

V. Jordan et al. (Eds.): HPCST 2020, CCIS 1304, pp. 80–94, 2020.
https://doi.org/10.1007/978-3-030-66895-2_6

1 Introduction

Intensive development of modern computation systems and technologies provides great opportunities for realization of multilevel and multiscale mathematical models. Owing to high-performance computation systems, allowing combining a great number of computing elements for synchronized work with common data, it becomes possible to study the parts and constructions using more precise and physically justified mathematical models. A possibility to deepen and to enhance mathematical models for the description of deformation and destruction of the materials and alloys, to include physical mechanisms at deeper scale levels up to atomic ones opens up.

Creation of mathematical models for the description of the behaviour of metal parts and constructions is a relevant task, which is determined by a large variety of the parts themselves and the modes of their operation, as well as by a great number of technologies of articles creation. The technologies for manufacturing parts, based on severe inelastic deformations, are widely used [1–10], providing substantial improvement of operating characteristics of the finished product [11–16]. A possibility of forecasting the moment of construction destruction and improvement of its reliability is very important [17]. Improvement of technologies for manufacturing parts and preliminary treatment of materials is impossible without development of mathematical models, describing the material behaviour in the course of manufacture of parts as experimental studies, despite the diversity of modern approaches and technological infrastructure, do not allow analysing all the physical mechanisms, participating in deformation and destruction (especially in the depth of solids). Besides, the conditions of laboratory experimental studies, usually, are not identical to the realised ones during the operation of parts, which does not allow to the fullest extent using the results, obtained on the samples, for forecasting the behaviour of real constructions [18, 19].

The wide-spread classical models, based on the macrophenomenological or structural mechanical approach, as well as experimental works, are unable to provide a researcher with complete information on the internal material structure. Frequently researchers use standard or original finite element packages, which allow monitoring the stress-strain behaviour of the entire construction at the macrolevel; however, the description of the evolution of the intragranular microstructure remains unavailable. At the same time, usually, such relevant factors as anisotropy of material properties are not considered, and the description of the physical mechanisms of deformation and destruction of the material has been simplified [20, 21].

One of the most promising approaches to the description of the processes of inelastic deformation of metals is an approach, based on the introduction of internal variables (IV) and application of crystal plasticity (CP). CP allow explicitly describing diverse physical mechanisms and effects at different scale levels: starting from the emergence, movement and interactions of dislocations at low scales and ending with the destruction of the sample under study or the entire construction. The models, based on CP, allow describing the evolution of the meso- and microstructure of the material and, as a consequence, by means of such models it is possible to control performance properties of the finished product [22]. Such opportunities allow creating so-called functional materials (materials, whose properties are in the best way suitable for a specific as-designed part).

The development of functional materials is particularly relevant for the creation of critical structures. As a rule, when creating functional materials, a boundary-value problem is solved, taking into account the matched parameters of the technological process for manufacturing a part; besides, it is evident that to solve this problem, it is necessary to use multilevel models [17, 23, 24].

The choice of the multilevel approach also is justified for the description of destruction processes. This process is very complex, multilevel and multistage; therefore, it is important to consider several structurally-scaled levels to simulate the emergence of micro-cracks and their subsequent development and proliferation, which can result in the formation of the main crack in the part [25].

The choice of the multilevel approach, based on CP, for the description of the processes of deformation and destruction of metal parts, is thus justified and preferable in the framework of this research. The implementation of multilevel models is a daunting task, which is complicated from the viewpoint of applied approaches and technologies of programming; however, this complexity is justified by the universality of the final software product, which can be supplemented with different embeddable modules, describing separate physical effects and peculiarities of the material or loading. In this connection, the basis of the software package is a method of finite elements, which allows simulating the loading of parts and constructions of practically any configurations.

2   Materials and Methods

2.1   Approach Description

Multilevel models, based on the introduction of internal variables, physical theories of elastoplasticity and continual consideration of the evolution of dislocations on slip systems, allow explicitly describing physical mechanisms and their carriers. It should be emphasized that the mathematical model is applied for the solution of boundary-value problems, thus analyzing the behavior of not the material, but of the specific construction, made according to quite a specific technology and modes. In this case, it is possible to analyze different conditions of loading of specific parts, as well as to consider different products, made of various materials with the replacement of appropriate internal variables and the geometry of the construction itself.

The process of deformation and destruction is multilevel and multistage. To take into account the realized deformation mechanisms, multilevel simulation is applied. An advantage of this approach is the introduction of internal variables (IV) [17, 26, 27], reflecting the state and the evolution of the material microstructure; besides, explicit IV enter the equations of state of the constitutive model. In most cases, the second group of IV (implicit IV) belongs to deeper scale levels and is included in the evolution equations as variables.

The developed mathematical model, taking into account the internal structure of the material, allows researching different parts and constructions under different conditions of loading, establishing their optimal properties and methods of obtaining the latter. In the future, this model will be useful in the development of functional materials as it allows tracing the change in the microstructure under certain conditions of loading.

2.2 Model Structure

From the viewpoint of scales, direct models are single-level [17]; at the same time, a subdivision into structural levels from the viewpoint of implementation of different physical mechanisms of material deformation is applied. In this way, the structure of the developed model has a level for the description of elastic and plastic deformation of crystallites (grains, subgrains, fragments), in the framework of which the variables are continual mechanical characteristics – stress, strain rates, strains. The explicit IV of this level can be a tensor of elastic properties, spin tensor, inelastic strain rate; the implicit IV include the shear rates along the slip systems (SS), orthogonal tensor, determining the orientation of crystallite lattices.

The next structural level is oriented to the description of the evolution of dislocation densities. The variables in the elements of this level are dislocation densities and dislocation motion rates; besides, such mechanisms are taken into consideration as nucleation of new dislocations, annihilation of dislocations of different signs on each of the SS, inflow and outflow of dislocations between the grain parts, formation of the aggregates and micro-cracks nucleation. The explicit IV of this level may include rates and densities of dislocations, shear rates, critical shear stresses on SS. The implicit IV include rates of change of the dislocation density owing to nucleation of new and annihilation of existing dislocations, barrier densities. Hence, the evolution of dislocation densities is considered at a lower structural level, but their contribution into the plastic deformation is determined by summing up the shears along all the slip systems.

A polycrystalline aggregate (macroscopic representative volume), consisting of a set of grains, is considered. Each of the grains is subdivided into the parts, for which all the relationships of the model are recorded. For the entire region of interest, a boundary-value problem of elasto-visco-plasticity is set and solved, within which the equilibrium equations in velocities, boundary and initial conditions are formulated [28]. To numerically implement the model on the whole, the finite-element method is used.

In the framework of the developed model, the deformation of the separate crystallites, including their rotations, the dislocation structure evolution, micro-cracks nucleation and their growth are studied; the relationships for each of the levels are given below.

2.3 Description of Elastic and Plastic Deformation

This submodel serves for the description of elastic and plastic deformation, as well as crystallites rotations. The input data for it are the deformation rates, determined from the solution of the boundary-value problem for the representative macrovolume, and shear rates along the slip systems, established from the submodels of lower structural levels. Using the indicated data at the level under consideration, elastic and plastic constituents of the deformation rate, stress and crystallites rotations are calculated.

A tensor of elastic properties Π is known, whose components are constants in the basis of the moving coordinates, connected to the crystallite lattice. Rotational speed are described by the spin tensor $\boldsymbol{\omega} = \dot{\mathbf{o}} \cdot \mathbf{o}^{\mathrm{T}}$, determining quasi-solid movement of the crystallite [17], where $\mathbf{o}$ is the orthogonal tensor connecting vectors of the basis of rigid moving and laboratory coordinates.

A hypothesis on additivity of elastic and inelastic constituents of the deformation rate measure $\mathbf{z} = \mathbf{z}^e + \mathbf{z}^{in}$ is accepted. The first summand describes an elastic constituent of the deformation rate and characterizes the lattice distortions; the second summand (plastic constituent) is determined by the shear rates along the slip systems and makes the lattice invariant. The basic constitutive relation at this level is the elastic law in the velocity relaxed form:

$$\dot{\sigma} + \sigma \cdot \omega - \omega \cdot \sigma = \Pi : (\mathbf{z} - \mathbf{z}^{in}), \tag{1}$$

where σ is the stress tensor, $\mathbf{z} = \hat{\nabla}\mathbf{v}^{\mathrm{T}} - \omega$ is the asymmetrical indifferent deformation rate measure, $\hat{\nabla}\mathbf{v}^{\mathrm{T}}$ is the transposed gradient of displacement velocity, determined in terms of the actual configuration.

The plastic deformation rate is determined by summing up the shear rates along all the slip systems:

$$\mathbf{z}^{in} = \sum_{k=1}^{n} \dot{\gamma}^{(k)} \mathbf{b}^{(k)} \mathbf{n}^{(k)}, \tag{2}$$

where $\dot{\gamma}^{(k)}$ is the shear rate in the k-th slip system, $\mathbf{b}^{(k)}$ and $\mathbf{n}^{(k)}$ are the unit vectors of the slip direction and normal's of the slip plane of the k-th slip system, respectively. The tangential (shear) stress, acting in the k-th slip system, is determined as follows:

$$\tau^{(k)} = \mathbf{b}^{(k)} \mathbf{n}^{(k)} : \sigma. \tag{3}$$

The condition of dislocation motion activation is achievement of some critical value by the tangential stress according to the Schmid's law [29]. This parameter is required when describing the dislocation motion and it provides consistency of equations.

2.4 Dislocation Motion Description

The dislocation-oriented submodel is used, in which scalar dislocation densities on slip systems are introduced as IV, for which appropriate evolution equations are formulated.

Edge dislocations are considered as majority carriers of plastic deformation; in addition, the subdivision into positive and negative dislocations, depending on the arrangement of extra-planes, is taken into account:

$$\rho^{(k)} = \rho_+^{(k)} + \rho_-^{(k)}. \tag{4}$$

It is supposed that dislocations are homogeneously distributed throughout the planes of each SS. Dislocations, independently of the sign, can be constrained by obstacles (barriers of dislocation and non-dislocation nature, grain boundaries, etc.), which is considered in the relationship for critical shear stresses. To activate the edge dislocation glide, it is necessary to apply tangential stresses, exceeding some critical value of stresses, which is determined by the chemical composition of the material, accumulated defects of different nature, etc.

The dislocation rate implies the average dislocation motion rate on the slip system. The dislocation rate depends on the temperature, at which deformation takes place, on the acting and critical shear stresses on the slip system and is determined by the following relationship [30, 31]:

$$v_{\pm}^{(k)} = \pm v_0^{(k)} \exp\left(\frac{-Q}{kT}\right) H\left(\left|\tau^{(k)}\right| - \tau_{c\pm}^{(k)}\right) sign(\tau^{(k)}), \tag{5}$$

where v_0 is the parameter, determined by the length of the dislocations free path, Q is the activation energy value, k is the Boltzmann constant, T is the absolute temperature, $\tau_{c\pm}^{(k)}$ are the critical shear stresses, $H(\cdot)$ is the Heaviside function. It should be noted that shear stress in the k-th slip system $\tau^{(k)}$ is determined by the stress of the overlaying level according to the Eq. (3).

Changes in the critical shear stresses are established taking into account the interactions of dislocations of different slip systems and the barrier influence. The additivity of different rates changing mechanisms the critical stresses is assumed. In a general view, the relationship for the change rates in critical stresses can be represented as follows:

$$\dot{\tau}_{c\pm}^{(k)} = \dot{\tau}_{dis\pm}^{(k)} + \dot{\tau}_{bar\pm}^{(k)}, \tag{6}$$

where

$$\dot{\tau}_{dis\pm}^{(k)} = \sum_{j=1}^{n} \frac{M^{(kj)}}{\sqrt{\rho^{(j)}}} \dot{\rho}^{(j)},$$

determines strengthening due to the influence of dislocation stress fields of different slip systems on dislocations in the existing slip system; $M^{(kj)}$ is the matrix of coefficients, describing the influence of dislocation stress fields, belonging to different slip systems, one upon another;

$$\dot{\tau}_{bar\pm}^{(k)} = \sum_{j=1}^{n} \frac{B^{(kj)}}{\sqrt{\rho_{bar}^{(j)}}} \dot{\rho}_{bar}^{(j)},$$

establishes strengthening due to the action of dislocation barriers of Lomer–Cottrell junction type [32], which cause hindering of dislocations, moving along two slip systems, on which the interaction of split dislocations has led to barrier formation. This hindering is determined by stress, caused by the barrier with respect to approaching dislocations. In this case, $\dot{\rho}_{bar}^{(j)}$ is the rate of changing dislocation barriers density, $B^{(kj)}$ is the matrix, describing the influence of stress fields of barriers of neighboring j-th SS on the moving dislocations of the k-th SS, respectively. The coefficients of matrices $M^{(kj)}$ and $B^{(kj)}$ have been calculated by solving separate subtasks. The dislocation (or the barrier) has been surrounded by dislocations at all SS one by one; then the stress, with which the located dislocation (or the barrier) influenced the surrounding dislocations, has been estimated. This is the way the coefficients of interaction of all possible SS pairs have been obtained.

The Orowan-type equation is used to determine the shear rate, taking into account the subdivision of dislocations into positive and negative ones:

$$\dot{\gamma}^{(k)} = \dot{\gamma}_+^{(k)} + \dot{\gamma}_-^{(k)} = b^{(k)}\left(\rho_+^{(k)} v_+^{(k)} - \rho_-^{(k)} v_-^{(k)}\right), \tag{7}$$

where $b^{(k)}$ is the Burgers vector. The dislocation rates are described in the coordinate system, which is common for positive and negative dislocations. The sign "minus" takes into account the circumstance that when the tangential stress is applied, negative and positive dislocations move in the opposite directions, but the contribution from the motion of these dislocations to the shear rate is similar in the sign. In this case, the values can differ. Later, the shear rate is used for the determination of inelastic deformations on each slip system and in each crystallite.

2.5 Description of Dislocation Densities Evolution

The dislocation density can change owing to such mechanisms as nucleation of new dislocations (Frank-Read sources), annihilation of dislocations of different signs on each of the slip systems, barriers formation. The relationship, describing the change in the dislocation densities, can be written in a general form as follows [33]:

$$\dot{\rho}_{\pm}^{(k)} = \dot{\rho}_{nuc\pm}^{(k)} - \dot{\rho}_{ann}^{(k)} - \dot{\rho}_{bar}^{(k)}. \tag{8}$$

For each of the summand of the relationship (8), an appropriate evolution equation has been formulated. The generation of new dislocations in the grain is conditioned by the work of the sources inside the crystallite. The dislocation generation rate is determined by the source density and actual stresses; the source is a dislocation segment, fixed with the impassable barrier. In a general view, the rate of change in the dislocation density owing to nucleation of new ones can be written in the following way:

$$\dot{\rho}_{nuc\pm}^{(k)} = k_{nuc}\rho_{src}^{(k)}f(\tau^{(k)}), \tag{9}$$

where k_{nuc} is the model parameter, $\rho_{src}^{(k)}$ is the active Frank-Read sources density on the k-th slip system, $f(\tau^{(k)})$ is the function, taking into account the difference of the current stress and critical stress of source activation.

The annihilation of dislocations with different signs on one or close parallel SS, when approaching each other up to the critical distance, is taken into account. The annihilation intensity depends on the density of dislocations and their motion rate [34]:

$$\dot{\rho}_{ann}^{(k)} = h_{ann}\rho_{+}^{(k)}\rho_{-}^{(k)}\left|v_{+}^{(k)} - v_{-}^{(k)}\right|, \tag{10}$$

where h_{ann} is the typical scale of dislocation annihilation; this parameter is used as a scaling factor.

Dislocations of some slip systems during interaction can form dislocation barriers, the type of which depends on a specific pair of slip systems and the value of stacking fault energy. The rate of change of the density of such barriers is determined by the following relationship (by the example of FCC with 12 slip systems):

$$\dot{\rho}_{bar}^{(k)} = k_{bar} \sum^{l=1,12/13,24} R_{bar}^{(kl)}\rho^{(k)}\rho^{(l)}\left\langle v^{(k)}\right\rangle, \tag{11}$$

where k_{bar} is the model parameter. Matrix R_{bar}, determining the type of the dislocation barrier, can be found from geometrical and physical considerations – by the possibility of formation and the energy of barrier formation on split dislocations (by the type of the crystal lattice and mutual arrangement of different slip systems) [32].

When using dislocation mechanisms for the description of the destruction, it is accepted that when reaching the critical density, an aggregate of dislocations of accumulation transforms into a micro-crack.

2.6 Transition from Deformation to Destruction

Micro-cracks nucleation is possible during the formation of the aggregate of dislocations of the critical length. To make the crack nucleation energetically favorable, it is necessary to save part of the work of the external forces in the form of latent energy or to create a region of great overstresses. It is supposed that it is possible to reach the critical state of the element, after which it is considered destructed. Achievement of the critical value of the dislocation density on one of the slip systems is accepted as one of the probable criteria of transition [35]. It is supposed that the crack forms in the destructed element; it loses its ability to resist loads and remains as such up to the completion of the loading process. After the destruction point, loading is redistributed among the neighboring elements, which can result in their further damage. The method of the description of the destructed state is the introduction of fictitious forces, which release the element with a micro-crack from stresses and resistance to influences.

3   Solving Algorithm

To solve the problem, a stepwise iteration procedure with the determination of the stress-strain behavior of crystallites and values of all internal variables is used. The first stage at each step of loading is obtaining the solution in rates for the representative macrovolume and then the integration of the systems of ordinary differential equations from the position of an observer in the moving coordinate system [17].

The considered element of the lower structural level will be a part of the crystallite; besides the volume, boundaries, orientation of the crystal lattice, as well as character-istics of all the neighbouring elements, are known. At the moment when the deforma-tion begins, all the parameters of the model, initial and boundary conditions have been defined. The conditions of loading are specified.

The first minor step should be emphasized owing to trivial stresses, when actual stresses (including tangential stresses on SS) are determined from the solution of the elastic problem; critical stresses on each SS and the grain orientation are established. After that, it is possible to pass to the description of some arbitrary step of loading. To begin with, the presence of micro-cracks in the element is checked, which may entail the loss by the element of the capability to perceive certain loadings.

The first stage is calculation of rates of change of implicit and explicit variables at the level of the description of the evolution of the dislocation structure. The rates of the dislocation density change are calculated owing to nucleation of new ones $\dot{\rho}_{nuc\pm}^{(k)}$ (determined by the relationship (9)), annihilation $\dot{\rho}_{ann}^{(k)}$ (10), barrier formation $\dot{\rho}_{bar}^{(k)}$ (11). Knowing these values, it is possible to find the rates of change of dislocation densities $\dot{\rho}_{\pm}^{(k)}$

according to (8). The rates of change of critical stresses $\dot{\tau}_{c\pm}^{(k)}$ are determined according to the Eq. (6); having substituted them in the relationship (5), the dislocation motion rates $v_{\pm}^{(k)}$ are obtained. Using integration, dislocation densities and critical stresses are obtained at the end of the step. Knowing the densities and average dislocation motions rates, it is possible to calculate the shear rates $\dot{\gamma}^{(k)}$ according to the Eq. (7), which are transferred to the overlaying level.

The actual tangential stress $\tau^{(k)}$ is calculated at the level of the description of elastic and plastic deformation using (3). It must be noted that this parameter is transferred to the level of the description of dislocation motion to calculate the rates. Knowing the shear rates, it is possible to calculate the rate of inelastic deformations $\mathbf{z}^{in}$ according to the Eq. (2). Later, the spin of the moving coordinates and the rate of inelastic deformations are determined. The obtained values are substituted into the Eq. (1), the integration of which leads to obtaining of actual stresses $\boldsymbol{\sigma}$.

Then let us proceed to the next step of loading and to the check of conditions of micro-crack nucleation. As it has been mentioned before, achievement of the critical value of dislocation densities on one of the slip systems is accepted as a possible criterion of micro-crack nucleation [35]. It is assumed that if the crack forms in the element, it loses an ability to perceive loadings and remains as such until the completion of the loading process. After the destruction point, the loading is redistributed among the neighboring elements, which may result in their further damage. The method of the description of the destructed state is the introduction of fictitious forces, which release the element with the micro-crack from stresses and resistance to influences.

The introduction of fictitious forces is reduced to the search for such nodal forces in the finite elements that reduce the actual stresses in the element to zero. Using the principle of virtual displacements (the work of internal stresses on virtual deformations equals the work of nodal forces on virtual displacements of nodes), the relationships for the fictitious forces have been obtained. Adding the found fictitious nodal forces with the opposite sign to the active forces, the problem is solved again at the same step. Hence, the region with trivial internal stresses in the destructed element is formed. This element remains in the calculation, it has certain rigidity; therefore, after additional loading at the next step of loading, stresses can appear in it again and the procedure of finding nodal forces is repeated.

4   Model Implementation

At the Department of Mathematical Modeling of Systems and Processes, Perm National Research Polytechnic University, a software package has been developed which is intended for the solution of problems of studying the processes of thermomechanical treatment of metals. The need for the creation of this software product is conditioned, first, by the relevance of solution of boundary-value problems of metal treatment in case of heavy gradients of displacements. Second, it is conditioned by the variety of the problems under consideration, whose solution implies numerical implementation, using basic and unique algorithms and approaches. In addition, the existing software packages have a limited functionality in case of the arbitrary model of the material and difficulties in using geometrically non-linear relationships. The developed software package allows removing these disadvantages; besides, there is a possibility of using parallel calculations.

In the created software package, a nucleus of the mathematical model has been realized, including the main constitutive relations. A calculating procedure has been implemented with the application of technologies of parallel programming, which accelerates calculations significantly. The system structure implies a possibility of its broadening, modification of the constitutive model, consideration of different mechanisms of deformation and physical effects (depending on the material structure and loading conditions). The accessibility of improvement is provided by the block structure of the system, realized in terms of object-oriented programming. To implement the system, the entities have been identified that correspond to objects of different structural and scale levels. The hierarchy of classes, corresponding to the mentioned levels, has been created. Two classes have been specified as base ones: crystallite and polycrystal. In the closed part of the base classes, there is a description of the hierarchy and the procedure of inheritance, main internal variables and evolutionary relationships, as well as realization of calculation logic. A possibility of redefinition of internal variables and modification of the relationships in the derived classes is provided.

To describe the evolution of the dislocation structure, its influence on the deformation and destruction of the material, a separate module has been developed, which is available for the introduction into the mentioned complex and for a possible combination with other subprograms. The model allows simulating the evolution of the dislocation density for the crystallite with various input data.

The method selected for integrating differential equations is the Euler's scheme of the first order with the fixed stride parameter by the time up to the value, the minimization of which does not influence significantly the result accuracy (for the considered examples $- \Delta t = 0.001$ s).

The module was implemented in IDEPyCharm 2019.2.6 (Community Edition) in the high-level interpreted language Python 3.7. The choice of the language is conditioned by the high velocity of code creation and a wide variety of alternate libraries, improving the performance and simplifying some calculations. For instance, the attached package NumPy, which supports multidimensional arrays and different functions for their processing, was used in this implementation. Classical methods of creation and processing of arrays (lists) in Python are inferior in the velocity to compiled languages; however, the use of NumPy allows eliminating this drawback. In particular, all the values of tangential and critical stresses, shears, rates and dislocation densities along all SS were stored in the arrays of the NumPy library. In addition to storage, the NumPy arrays accelerated the performance of some tensor operations, such as calculation by the relationship (3). An interpreter PyPy was also used to accelerate the calculations; it interprets the initial Python-code into a language of a lower level (C-code), which leads to a significant optimization of the program execution time. Hence, a compromise between the code transparency and the computation speed has been successfully reached.

During execution of the program, the values of all crucial quantities were obtained in order to further draw graphs for the analysis of the results. In addition, IDEPyCharm 2019.2.6 allowed debugging efficiently the developed module.

5 Results

Elastoplastic deformation of the monocrystalline copper specimen was simulated; a tensor of the strain rate was specified as loading ($z_{12} = z_{21} = 5 \times 10^{-4}$ s^{-1}, the rest of the components equal zero). The specimen was located in such a manner that the crystallographic coordinate system coincided with the laboratory one. The parameters that are required for the calculation, such as Burgers vectors, normal vectors for SS, lattice parameter, tensor components of elastic properties are known from the literature. Deformation was carried out till reaching the intensity of accumulated deformation equal to 40%. At each moment of loading, the evolution of dislocation densities owing to such mechanisms as annihilation and nucleation of new dislocations was analyzed.

It has been established that 6 among 12 SS are active; the evolution of dislocation densities on these systems is presented in Fig. 1.

It is possible to notice that with minor strains, the process of changing the dislocation density owing to nucleation of new dislocations prevails over the change owing to annihilation. When entering the region of elasto-plastic deformation, a sharp increase in the dislocation density is observed. This is conditioned by the activation of the Frank-Read sources, which resulted from the growth of shear stresses on slip systems and the dislocation motion rate. The situation changes when cumulative strain increases. During loading, the average dislocation rate increases, which entails the intensification of the annihilation and deceleration of the dislocation density growth. The appearance of the obtained dependencies corresponds qualitatively to the known experimental data.

Besides, several computing experiments with the similar problem statement were conducted in conditions of loading, mentioned above, with different shear rates. During deformation, the evolution of dislocation densities was considered, and when the

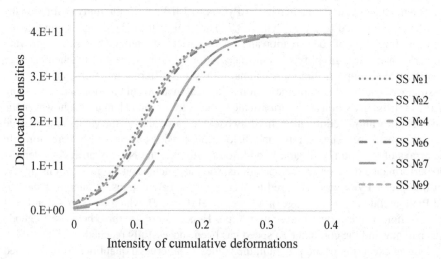

Fig. 1. Dependence of dislocation density on different slip systems on the strain intensity. SS designations: No. 1 **n** $= (1; 1; 1)$ **b** $= (1; 0; -1)$, No. 2 **n** $= (1; 1; 1)$ **b** $= (0; 1; -1)$, No. 4 **n** $= (-1; 1; 1)$ **b** $= (1; 1; 0)$, No. 6 **n** $= (-1; 1; 1)$ **b** $= (0; 1; -1)$, No. 7 **n** $= (1; -1; 1)$ **b** $= (1; 1; 0)$, No. 9 **n** $= (1; -1; 1)$ **b** $= (1; 0; -1)$.

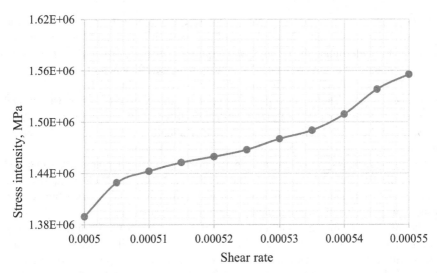

Fig. 2. Stress intensity at the moment of micro-crack nucleation with different values of the applied strain rate.

dislocation density on one of SS reached a critical value (formation of dislocation accumulation of critical capacity), nucleation of the micro-crack occurred on the indicated SS. As soon as this happens, the element is considered destructed and further calculations are terminated. In each separate computing experiment, the stress intensity was measured at the moment of micro-cracks formation (Fig. 2).

It has been shown that when the shear rate increases, the value of stress intensity at the moment of crack formation also reaches greater values. This is explained by the fact that at a lower shear rate, the material has enough time to relax due to the evolution of the dislocation structure. When the shear rate increases, the system has no time to relax, and the moment of the possible micro-crack formation occurs in case of a greater intensity of stresses.

In this way, during simple shear deformation, the evolution of dislocation densities owing to such mechanisms as new dislocation nucleation and annihilation was studied. In future, the influence on the change in the dislocation density during deformation of such mechanisms as barrier formation and initiation of dislocation flows between neighboring crystallites is planned to research. The mentioned mechanisms will be further taken into account in the solution of the boundary-value problem with the aim to analyze the evolution of the dislocation structure for the grain agglomeration (of a polycrystalline aggregate).

6   Conclusion

The use of numerical experiments provides great opportunities for the research of deformation and destruction processes of metals and alloys. In the work, different approaches

to the description of inelastic deformation and destruction of metal parts and constructions have been considered. The relevance of the research in this field has been empha-
sised, which is determined by the increased requirements to performance characteristics
of the manufactured metal products. A conclusion on the efficiency of using multilevel
models, based on crystal plasticity, including for the designing of functional materials, has been made. The detailed description of the direct elasto-visco-plastic model,
based on this approach, including the relations of different structural levels, a connection between them has been presented. Particular attention is given to the dislocation
dynamics, including a transition to the micro-crack nucleation.

The software package, implementing mathematical multilevel models of the material,
based on the finite-element approach and crystal plasticity, has been discussed. This
package can be expanded and supplemented with different modules, describing certain
physical mechanisms of different metals and alloys deformation.

The model implementation algorithm has been given: starting from the boundary-
value problem solution at the level of mono- and polycrystalline macro-sample and
ending with obtaining the characteristics of the dislocation structure. To describe the
element destruction and further spread of cracks, it is proposed to use fictitious forces.
Using them, it is possible to describe the loss by the element of the ability to resist
loading, without removing it from the system and the computational grid.

The results, obtained by means of the numerical implementation of the submodel for
the description of the dislocation densities evolution, have been given. The simulation
was carried out in IDE PyCharm 2019.2.6 (Community Edition) in Python 3.7, using the
computation process acceleration. Elasto-plastic deformation of the copper crystallite
has been simulated; at the same time, particular attention has been given to physical
mechanisms, influencing the dislocation density. The obtained dependencies of the total
dislocation density on the intensity of the accumulated strain comply qualitatively with
the known experimental data. The sequential activation of nucleation and annihilation
mechanisms of dislocations on active SS has been shown. The research results of the
dependence of the stresses intensity, under which the dislocation density on one of the
SS reached the critical value, on the deformation rate have been presented.

The developed software product is a basis for further research in the field of micro-
cracks nucleation in metals and has a theoretical and practical value.

Acknowledgments. The work was supported by the Russian Science Foundation (grant No.
17-19-01292).

References

1. Bengus, V.Z., et al.: Plasticity of the nanostructural and polycrystalline titan at temperatures
   300, 77, and 4.2. Metallophys. Adv. Technol. **26**(11), 1483–1492 (2004). (in Russian)
2. Valiev, R.Z., Najmark, O.B.: Bulk nanostructured materials: unique properties and innovative
   potential. Innovations **12**(110), 70–76 (2007). (in Russian)
3. Gorynin, I.V.: Creation of structural and functional nanomaterials. Innovations **6**(116), 34–43
   (2008). (in Russian)

4. Islamgaliev, R.K., Nesterov, K.M., Khafizova, E.D., Ganeev, A.V., Golubovskiy, E.R., Volkov, M.E.: Strength and fatigue of ultrafine-grained aluminum-based alloy AK4-1. Vestnik UGATU **16**(8–53), 104–109 (2012). (in Russian)

5. Islamgaliev, R.K., Ganeev, A.V., Nikitina, M.A., Karavaeva, M.V.: Structure and properties of ultrafine-grained martensitic steel. Vestnik UGATU **20**(3–73), 19–24 (2016). (in Russian)

6. Sitdikov, O.S.: Effect of multidirectional forging on the fine-grained structure development in a high-strength aluminum alloy. Lett. Mater. **3**(3), 215–220 (2013). (in Russian)

7. Gleiter, H.: Nanocrystallinematerials. Prog. Mater. Sci. **33**, 223–315 (1989). https://doi.org/10.1016/0079-6425(89)90001-7

8. López-Chipres, E., Garcia-Sanchez, E., Ortiz-Cuellar, E., Hernandez-Rodriguez, M.A.L., Colás, R.: Optimization of the severe plastic deformation processes for the grain refinement of Al6060 alloy using 3D FEM analysis. J. Mater. Eng. Perform. 1–7 (2010). https://doi.org/10.1007/s11665-010-9783-1

9. Suwas, S., Bhowmik, A., Biswas, S.: Ultra-fine grain materials by severe plastic deformation: application to steels. In: Haldar, A., Suwas, S., Bhattacharjee D. (eds.). Microstructure and Texture in Steels, pp. 325–344 (2009). https://doi.org/10.1007/978-1-84882-454-6_19

10. Valiev, R.Z., Islamgaliev, R.K., Alexandrov, I.V.: Bulk nanostructured materials from severe plastic deformation. Prog. Mater. Sci. **45**(2), 103–189 (2000). https://doi.org/10.1016/s0079-6425(99)00007-9

11. Valiev, R.Z., Aleksandrov, I.V.: Nanostructured materials obtained by severe plastic deformation. Logos, Moscow (2000). (in Russian)

12. Noskova, N.I., Mulyukov, R.R.: Submicrocrystalline and nanocrystalline metals and alloys. UrO RAN, Ekaterinburg (2003). (in Russian)

13. Murashkin, M.Y., Markushev, M.V., Ivanisenko, Y.V., Valiev, R.Z.: Strength of commercial aluminum alloys after equal channel angular pressing (ECAP) and post-ECAP processing. Solid State Phenom. **114**, 91–96 (2016). https://doi.org/10.4028/www.scientific.net/SSP.114.91

14. Hong, X., Godfrey, A., Zhang, C.L., Liu, W., Chapuis, A.: Investigation of grain subdivision at very low plastic strains in a magnesium alloy. Mater. Sci. Eng. **693**, 14–21 (2017). https://doi.org/10.1016/j.msea.2017.03.080

15. Dobatkin, S., et al.: Grain refinement, texture, and mechanical properties of a magnesium alloy after radial-shear rolling. J. Alloy. Compd. **774**, 969–979 (2019). https://doi.org/10.1016/j.jallcom.2018.09.065

16. Song, M., et al.: Grain refinement mechanisms and strength-hardness correlation of ultrafine grained grade 91 steel processed by equal channel angular extrusion. Int. J. Pressure Vessels Pip. **172**, 212–219 (2019). https://doi.org/10.1016/j.ijpvp.2019.03.025

17. Trusov, P.V., Shveykin, A.I.: Multilevel models of mono- and polycrystalline materials: theory, algorithms, application examples. SO RAN, Novosibirsk (2019). https://doi.org/10.15372/multilevel2019tpv (in Russian)

18. Volegov, P.S., Gribov, D.S., Trusov, P.V.: Damage and fracture: review of experimental studies. Phys. Mesomech. **19**(3), 319–331 (2016). https://doi.org/10.1134/s1029959916030103

19. Tang, X.S., Wei, T.T.: Microscopic inhomogeneity coupled with macroscopic homogeneity: A localized zone of energy density for fatigue crack growth. Int. J. Fatigue **70**, 270–277 (2015). https://doi.org/10.1016/j.ijfatigue.2014.10.003

20. Naimark, O.B.: Collective properties of defect ensembles and some nonlinear problems of plasticity and fracture. Phys. Mesomech. **6**(4), 45–72 (2003). (in Russian)

21. Volegov, P.S., Gribov, D.S., Trusov, P.V.: Damage and fracture: classical continuum theories. Phys. Mesomech. **20**(2), 157–173 (2017). https://doi.org/10.1134/s1029959917020060

22. Volegov, P.S., Gribov, D.S., Trusov, P.V.: Damage and fracture: crystal plasticity models. Phys. Mesomech. **20**(2), 174–184 (2017). https://doi.org/10.1134/s1029959917020072

23. McDowell, D.L., Olson, G.B.: Concurrent design of hierarchical materials and structures. Sci. Model Simul. **15**(1–3), 207–240 (2010). https://doi.org/10.1007/s10820-008-9100-6

24. Nakamachi, E., Kuramae, H., Sakamoto, H., Morimoto, H.: Process metallurgy design of aluminum alloy sheet rolling by using two-scale finite element analysis and optimization algorithm. Int. J. Mech. Sci. **52**, 146–157 (2010). https://doi.org/10.1016/j.proeng.2011.04.372

25. Vladimirov, V.I.: The physical nature of metals fracture. Metallurgy, Moscow (1984). (in Russian)

26. Gerard, A.: Maugin Continuum Mechanics of Electromagnetic Solids, North-Holland etc, (1988)

27. McDowell, D.L.: Internal state variable theory. In: Yip, S. (ed.), Handbook of Materials Modeling. Springer, pp. 1151–1169 (2005). https://doi.org/10.1007/978-1-4020-3286-8_58

28. Pozdeev, A.A., Trusov, P.V., Nyashin, YuI: The Large Elastoplastic Deformation: Theory, Algorithms, Applications. Nauka, Moscow (1986). (in Russian)

29. Taylor, G.I.: Plastic strain in metals. J. Inst. Metals **62**, 307–324 (1938)

30. Meyers, M.A., Benson, D.J., Vohringer, O., Kad, B.K., Xue, Q., Fu, H.-H.: Constitutive description of dynamic deformation: physically-based mechanisms. Mater. Sci. Eng. **322**(1–2), 194–216 (2002)

31. Roters, F., Eisenlohr, P., Hantcherli, L., Tjahjanto, D.D., Bieler, T.R., Raabe, D.: Overview of constitutive laws, kinematics, homogenization and multiscale methods in crystal plasticity finite-element modeling: theory, experiments, applications. Acta Materialia **58**, 1152–1211 (2010). https://doi.org/10.1016/j.actamat.2009.10.058

32. Hirth, J.P., Lothe, J.: Theory Of Dislocations. McGraw-Hill, New York (1968)

33. Austin, R.A., McDowell, D.L.: A dislocation-based constitutive model for viscoplastic deformation of FCC metals at very high strain rates. Int. J. Plast. **27**(1), 1–24 (2011)

34. Leung, H.S., Leung, P.S.S., Cheng, B., Ngan, A.H.W.: A new dislocation-density-function dynamics scheme for computational crystal plasticity by explicit consideration of dislocation elastic interactions. Int. J. Plast. **67**, 1–25 (2015). https://doi.org/10.1016/j.ijplas.2014.09.009

35. Stroh, A.N.: The formation of cracks as a result of plastic flow. Proc. R. Soc. London. Ser. **223**, 404–414 (1954). https://doi.org/10.1098/rspa.1954.0124

The Three-Level Model to Describe Serrated Yielding: Structure, Algorithm, Implementation, Results

Dmitriy Gribov$^{(\boxtimes)}$ (ID), Fedor Popov (ID), and Eugenia Chechulina (ID)

Perm National Research Polytechnic University, 614990 Perm, Russian Federation
gribowdmtrii@yandex.ru, popovfyodor@yandex.ru,
zhenya-chechulina@yandex.ru

Abstract. Modeling the processes of plastic deformation is a relevant problem of Solid State Physics which, in reality, requires using advanced computational facilities. A mathematical model development is followed by building an algorithm for this model's implementation, the model implementation itself and numerical experiments. Numerical experiments can be used both to describe deformation processes, to identify and verify a model's parameters. It is possible to use parallel computations to implement models built on hierarchical principles. The paper aims at describing serrated yielding (the Portevin–Le Chatelier effect). By analyzing the physical foundations of the process, we utilize the hypothesis about the occurrence of serrated yielding due to interaction of impurity atoms and edge dislocations. During plastic deformation, the impurity atoms are captured by moving edge dislocations; as the dislocation velocity increases due to increasing stresses, clouds of the impurity atoms are detached from the dislocations, which ultimately leads to a jump-like behavior of the stress-strain curve. A submodel describing the interaction of the impurity atoms with dislocations is built into a three-level model of viscoelastic deformation, which describes the material's behavior at the representative volume. The model makes it possible to describe main evolution mechanisms of the dislocation substructure. Relations are given to determine the interaction forces of the impurity atoms with the dislocations, the hardening law, which takes into account the number of the impurity atoms fixed on the dislocations and their current density. The general model structure, relations between the submodel parameters of different levels are considered, the algorithm description of the model implementation is presented. The model algorithm is implemented using a program in the c++ language with blocks of parallel execution, the analysis of the program's performance is made. By using the implemented program we carried out a number of numerical experiments, such as the macrosample deformation, the evolution analysis of dislocation densities on slip systems, the process intensity of catching and releasing the impurity atoms from the slip systems, and the dependence of stress intensities on strain intensities.

Keywords: Multilevel model · Implementation algorithm · The Portevin - Le Chatelier effect · Numerical experiments

© Springer Nature Switzerland AG 2020
V. Jordan et al. (Eds.): HPCST 2020, CCIS 1304, pp. 95–113, 2020.
https://doi.org/10.1007/978-3-030-66895-2_7

1 Introduction

The description of metal deformation processes based on the approaches of crystal plasticity requires a large number of numerical experiments. Some experiments are necessary to determine a large number of parameters of the developed models. When we find them, it is possible to solve applied problems, i.e. determine the mechanical response of materials and structures. It should also be noted that multilevel models have recently been developed to describe inelastic deformation of metals. Due to the multilevel description, a more detailed process description is possible, as the built hierarchical system with large scale elements contains those of smaller scales, and enables a physically-based description of the material structure, provided that the relations between the levels are correctly described. The disadvantages of the multilevel models include massive computing resources for the implementation of numerical experiments. This fact reveals the need to optimize the program code, to implement the parallel execution of the model algorithm in the areas where it does not contradict with the numerical schemes and the structure of the mathematical model. At the same time, the studies including the numerical implementation of the multilevel models may unlock a deeper understanding of the studied area by being not only a substitute for natural experiments but also offering predictive opportunities.

The phenomenon of serrated yielding (the Portevin–Le Chatelier (PLC) effect) as a manifestation of the instability of inelastic deformation has been found for a number of metals and alloys in certain ranges of strain rates and temperatures under different loading programs. On deformation curves, serrated yielding manifests itself in the form of repetitive inhomogeneities, i.e. steps or teeth of various types depending on a type of loading. The localization laws of plastic deformation in a sample under loading, as noted in a number of works [1–5], and often correlate with the discontinuous nature of the deformation curve, with each jump in the uniaxial loading diagram (stress σ - deformation ε) associated with the emergence or propagation of a deformation band visible on the lateral surface of the sample.

The aim of studying the PLC effect when it comes to deformation of metals and alloys is to predict how temperatures and deformations influence behaviors of structures. Most structural metals and alloys in certain temperature-strain rate ranges of deformation demonstrate the manifestation of the PLC effect. Aluminum alloys have been recently used for critical structures and their parts operating in extreme conditions, as they own high strength characteristics but are subjected to the PLC effect. In this regard, it is a priority to solve the problem of establishing the ranges within which serrated yielding is realized, in order to exclude them from the list of manufacturing modes of processing metal products (especially at finishing operations).

The results of numerous field experiments on serrated yielding allow us to conclude that a discontinuous, irregular response of the material under monotonic impacts is caused by the presence of inhomogeneities in material properties and features of motion of large dislocation arrays (i.e., their coordinated motions) at various scale levels. These inhomogeneities can develop in a cascade manner from nanosizes to values comparable with the sizes of macrosamples.

The processes of inelastic deformation and properties of polycrystalline materials at the macrolevel are largely determined by the state of the evolving meso- and microstructure of a material. Dynamic strain aging (SDS) associated with the effect of the agglomeration of impurities near dislocations is assumed to be the major mechanism resulting in the PLC effect. The stress fields from point defects interact with the stress fields of dislocations, hindering the movement of dislocations until the current barrier (critical stress) is overcome, at which the dislocations are detached from the clouds of the impurity atoms, which is the main factor of the occurring serrated area on the deformation curve.

To describe the dynamic strain aging and the effects associated with it, most researchers (A. Benallal, A. Bertram, T. Böhlke, X. Chen, Y. Estrin, S. Graff, P. Hähner, O. Hopperstad, S.Y. Hu, L.P. Kubin, R. Larsson, C.P. Ling, P.G. McCormick, E. Rizzi, S. Zhang, S.-Y. Yang, D. Yu, etc.) [1–5] use macrophenomenological models of inelastic deformation [6].

L.P. Kubin and Y. Estrin [7] proposed one of the well-known phenomenological viscoplastic models, which takes into account strain aging and focuses on describing the PLC effect under uniaxial loading. The strain rate sensitivity of the material is determined by the sum of two parts, i.e. the one independent and dependent on deformation aging. The latter is established using the dislocation submodel, which includes evolutionary equations for scalar densities of mobile and forest (immobile) dislocations. It is noted that the model makes it possible to determine the critical values of accumulated strains (of the beginning and end of serrated yielding) that are actually observed in the experiments.

The macrophenomenological model proposed in [8] is based on a qualitative analysis of the physical processes responsible for the start of serrated yielding. The model is focused on describing uniaxial loading. An evolutionary equation for the flow stress is proposed which includes the terms responsible for strain and rate hardening, as well as for the change in flow stress due to strain aging. The latter takes into account the competition between hardening due to an interaction of the impurity atoms with dislocations and softening due to a separation of dislocations from the atmospheres.

The paper in [9] describes the macrophenomenological elastoviscoplastic model and the results obtained with its help in studying the deformation of specimens cut at different angles (related to the direction of rolling) from rolled sheets of aluminum alloy AA5083-H116. The constitutive relations are based on the McCormick model [10], in which the flow stress, along with the terms characterizing the strain and rate hardening, also includes a term describing the change in the flow stress with time due to strain aging.

Macrophenomenological models based on the macrosample experiments set dependences between the parameters of the macrolevel, without in-depth issues of a material microstructure evolution. In particular, the disadvantage of these models is that they are not suitable for predicting properties of designed materials. Also they do not make it possible to determine the main parameters of unstable deformation (width, propagation rate and value of deformation in localization bands). It is necessary to study the material behavior at structural-scale levels lower than the macrolevel to get a correct description of plastic deformation and its inhomogeneity, to consider the most significant physical mechanisms that determine it and accompany it.

Models based on crystal plasticity (J. Alcala, H. Askari, R.A. Austin, D.L. McDowell, J. Balik, P. Lukac, M.S. Bharathi, etc.) do not include disadvantages of the macrophe-nomenological models, which are based on an explicit consideration of deformation mechanisms at meso- and microscales.

The work in [11] suggests using a 2-level elastoplastic model to study the PLC effect. In order to describe the PLC effect, an evolutionary equation for critical stresses was formulated, which includes additive terms responsible for a change in the resistance to deformation due to athermal (strong) barriers, strain hardening (which also takes into account softening by dynamic recovery, rate hardening (excluding dynamic aging), and a separate term reflecting the effects of strain aging. The results using the specified elastoviscoplastic physical model to analyze the deformation of flat specimens from the Al-2.5% Mg alloy are presented.

The paper in [12] presents a constitutive plastic model describing the behavior of polycrystalline heat-resistant nickel-based alloys. Three scale levels are considered, the macrolevel corresponding to the representative volume of the polycrystal, the mesoscale (or the level of an individual grain), and the level of subgrains. The model studies monotonic loading. At the mesoscale, the structure of the nickel-based alloy is modeled using an additional plastic submodel using the finite element method.

A model proposed in [13] to determine the resistance to deformation is presented in the form of a simultaneous integro-differential equations describing the material hardening as a result of an increment in dislocation density and blocking of the motion of mobile dislocations by grain and subgrain boundaries taking into account softening due to recovery and recrystallization. It is proposed to describe the physical processes of hardening and softening, which determine the physical and mechanical properties of alloys by changing the shape of the deformation curve. To determine the resistance to deformation, additional internal parameters were introduced into the model, which are responsible for blocking the motion of mobile dislocations by the impurity atoms and inclusions. The model allows one to describe the resistance to deformation of alloys under high-temperature deformation taking into account the combined effect of hardening, recovery, recrystallization and barrier effects (blocking of free dislocations by the impurity atoms) as well as the Portevin–Le Chatelier effect. The parameters of the model were determined by identifying them according to the experimental data obtained under uniaxial tension of the specimens from the AMg6 alloy in the range of deformation rates from 5 to 25 s^{-1} and temperatures of 400 and 500 °C.

Some authors [14, 15] offer a review of theoretical works based on crystal plasticity and a description of deformation properties of alloys in temperature-rate ranges, in which diffusion processes have a significant effect on the material behavior. Particular attention is paid to the Portevin–Le Chatelier (PLC) effect. The disadvantages of models based on crystal plasticity include the problem of closing the set of relations, a large number of parameters and difficulties in identifying them, and the complexity of their numerical implementation.

1.1 Interaction of Impurity Atoms and Dislocations

It is known that the motion of dislocations is not continuous; it has a jump-like nature, while free motion alternates with stops at localized obstacles during a certain waiting time [16]. As the processes of inelastic deformation at the macrolevel are largely determined by the state of the evolving meso- and microstructure of the material, which may finally lead to a spontaneous deformation localization, both Russian and foreign researchers attempted to construct mathematical models by taking into account the dislocation dynamics (A. Alankar, A. Arsenlis, K.M. Davoudi, L. Kiely, D. Li, H. Zbib, C. Reuber, M. Zecevic, M. Knezevic, M. Lebedkin et al.).

A big part of the monograph in [17] refers to analyzing the interaction of impurities with dislocations. The impurity atoms move to the immobile dislocations by diffusion and are located along the dislocation lines in positions with the maximum binding forces. For mobile dislocations, two options are possible. Firstly, it refers to stresses exceeding a certain limit, in this case, dislocations can detach from the atmospheres formed around them. Secondly, it refers to low stresses, thus low velocities of motion, when dislocations can carry the atmospheres of the impurity atoms with them.

In [18], along with the inflow of atoms to a dislocation due to the interaction of stress fields, diffusion is taken into account due to the concentration gradient of the impurity atoms. To consider the processes over long periods of time, the author introduces the concept of the so-called average capture radius. It is the radius of a cylindrical region, which (in case of a specific problem) has the same probability of capturing the diffusing atoms just like the dislocation with its inherent stress field. Analytical solutions are obtained to determine the temporal and spatial evolution of the impurity atoms' concentration in the neighborhood of the single edge and screw dislocations. An analytical solution is obtained for the fraction of the deposited impurity atoms on an array of the parallel dislocations, which is described by a single dislocation and a cylindrical region with an average capture radius.

A detailed consideration of the interaction of point defects with dislocations is given in [19] with a brief overview of earlier works in this area. A variety of interactions of the point defects with the dislocations is noted, i.e. those determined by the stress fields of the dislocations during the introduction of defects with sizes different from lattice atoms, chemical and electrical effects, and vibrations of the defects. It is noted that the interaction of the point defects with the dislocations due to their own stress fields is the most important one.

A model based on considering the interaction of the dislocations with the impurity atoms was proposed in [20–22]. It is assumed that the flow stress is equal to the sum of the resistance to deformation from the interaction of the dislocations with the forest dislocations and the friction stress from the interaction of the dislocations with the impurity atoms, which depends on the concentration of the latter. It is assumed that the dislocations in motion can capture the impurity atoms and move together with them. The captured impurity atoms are capable of intensively diffusing along the lines of the dislocation nuclei up to a certain saturation limit. The dependences of the average velocities of motion of the dislocations and stress of resistance to flow on the concentration of the impurities and temperature are obtained; on the basis of which the existence of a temperature interval is shown where the negative sensitivity of the resistance to deformation on the strain rate takes place.

2 Concept and Mathematical Setting

In this work we used a model based on crystal plasticity (CP). The model was built using an approach based on the introduction of internal variables (internal parameters of the system state). The material's reaction at each instant of time is completely determined by the values of the tensor thermomechanical characteristics of the material and a finite set of internal variables, parameters of physical and mechanical effects and their time derivatives of the required order at the instant of time under study.

Some of the internal variables included in the constitutive relations of the level under consideration are called explicit internal variables. They include parameters characterizing yield stress, etc. The second group of the internal variables is an implicit one, that is, characterizing the evolution of the material structure and processes at lower scale levels of the model. By analyzing the existing material models and physical mechanisms of the inelastic deformation of a wide range of structural materials, it becomes possible to propose a constitutive model structure which includes constitutive relations, evolutionary and closing equations.

The positive aspects of using this approach include a clearer physical interpretation of the internal variables, the possibility of comparing it with the real material structure, great possibilities of processing the obtained data, as well as the fact that it is based on the fundamental physical laws [9].

2.1 General Structure of the Model

This work considers the macrolevel, meso-1, and meso-2 as structural-scale levels to describe the behavior of the polycrystalline specimen. The key mechanism of the inelastic deformation in this work is assumed to be the motion of the edge dislocations. At the mesoscale-2, its description is carried out using the introduction of the dislocation densities on slip systems (SS) and their velocities. At mesoscale-1, the inelastic deformation is considered in terms of shear rates along the crystallographic SS. To connect the submodels of different levels, the explicit internal variables are introduced into the structure of the constitutive relations at each of the scale levels, which are determined by the closing equations describing the deformation processes at deeper scale levels with respect to the considered one. At the macrolevel, the behavior of the material representative volume is described in terms of the elastic and plastic components of the strain tensor and stress tensor. The element of the macrolevel (macropoint, representative macrovolume) is hierarchically senior in relation to a certain set of elements of the meso-1. The hierarchical structure of the model makes it possible to simultaneously calculate the stress-strain state in all the elements of the underlying level, have a further volume averaging. This fact allows using parallel execution schemes when implementing this model. As an effect on the mesolevel-1 from the macrolevel, kinematic effects are transmitted (Voigt's hypothesis about the equality of the velocity gradients of the displacements of the two levels is accepted). A macropoint can be used to solve the boundary value problem; the inelastic component of the strain rate tensor of the macrolevel is determined by averaging the rates of inelastic deformations of the meso-1 elements. In describing the deformation of a crystallite, the mesoscale-1 model and meso-2 submodel are used. At the meso-1, the mechanical state of the crystallite is described. The tangential stresses on the slip systems determined at the mesoscale-1 are transmitted as an action on the mesoscale-2, where the evolution of defect densities is considered, changes in the concentration fields of the impurity atoms are calculated, hardening is described, the velocities of dislocations motion are determined, as well as shear rates along the SS. From the mesoscale-2 to the mesoscale-1, shear rates are transmitted along the slip systems.

The problem of describing the behavior of alloys under loading taking into account the evolution of the microstructure can be subdivided into several interconnected (sub) problems, namely, the problem of determining the stress-strain state (SSS), the problem of thermal conductivity, the problem of diffusion of the impurity atoms, i.e., the related problem splits according to the physical processes. To simplify the model, an assumption is introduced that the process under consideration is isothermal.

The mesoscale-2 model uses the distribution of the dislocation densities for the slip systems into positive and negative (according to a predetermined rule, depending on the direction of the extraplane). The mesoscale-2 variables are dislocation densities (positive and negative), the density of the dislocation sources on the SS, and the concentration of the impurity atoms for each SS. Internal variables of the mesoscale-1 include shear rates along the slip systems and critical stresses, which are determined by the dislocation substructure with allowance for interactions with the impurity atoms at the mesoscale-2. In the mesoscale-2 model, in addition to the interaction of dislocations with each other, the interaction of the dislocations with the impurity atoms is introduced, and the process of redistributing the edge dislocations over the SS is described.

The fundamental hypothesis in this work is a detailed physically justified description of the dislocation slip, the interaction of dislocations with each other, as well as the interaction of the dislocations with the impurity atoms. The latter makes a significant contribution to the motion of the dislocations due to the formation of clouds of the impurity atoms. Depending on the dislocation velocities, the clouds of the impurity atoms change in the volume and the contribution of the impurity atoms to the critical shear stresses along the SS changes.

3 Mathematical Setting

The strain rates are transferred from the macrolevel to the meso-1 level using the Voigt hypothesis:

$$\mathbf{z} = \mathbf{Z}. \tag{1}$$

At the mesolevel-1 we determine the plastic part of the strain rate measure $\mathbf{z}^{in}$, σ stress tensor components, the acting shear stresses being found $\tau^{(k)}$ are transmitted from the mesolevel-1 to the mesolevel-2. Shear rates $\dot{\gamma}^{(k)}$ are found at the mesolevel-2 by the velocities of the motion of the dislocations and are used to calculate the inelastic component of the measure of the strain rate at the mesolevel-1. The submodel of the mesolevel-1 includes the following system of equations:

$$
\begin{cases}
\dot{\sigma} + \sigma \cdot \omega - \omega \cdot \sigma = \pi : (\mathbf{z} - \mathbf{z}^{in} - \omega), \\
\mathbf{z} = \mathbf{z}^e + \mathbf{z}^{in} \\
\omega = \dot{\mathbf{o}} \cdot \mathbf{o}^T \\
\mathbf{z}^{in} = \sum_{k=1}^{n} \dot{\gamma}^{(k)} \mathbf{b}^{(k)} \mathbf{n}^{(k)} \\
\tau^{(k)} = \mathbf{b}^{(k)} \mathbf{n}^{(k)} : \sigma
\end{cases}
\tag{2}
$$

where $\mathbf{b}$ is the unit vector in the direction of the Burgers vector of the edge dislocation, $\mathbf{n}$ is the normal to the plane of the dislocation slip, k is the number of the slip system (SS), ω is the lattice spin. The description of the evolution of the microstructure is made at the mesoscale-2. The explicit internal variables (IV) of the mesoscale-1 include σ, the stress tensor at the mesolevel-1.

The evolutionary equations for the dislocation densities on the SS describe the nucleation of the dislocations due to the operation of the Frank-Read sources, the annihilation of the dislocations of different signs on one SS. The critical stresses on the SS are determined by the current density of defects on all the systems. The explicit internal variables at the mesoscale-2 include the average shear rates of $V_{+-}^{'(k)}$ dislocations depending on the tangential stresses τ, critical shear stresses τ_c and temperature θ, the density of the positive and negative dislocations on SS $\rho_+^{(k)}$, $\rho_-^{(k)}$.

The average velocities of dislocation motion are determined by the acting stresses and defect densities. Shear rates are determined using Orowan's velocity form equation. The meso-2 submodel is represented by the following system of equations:

$$
\begin{cases}
V_{\pm}^{'(k)} = \pm f_1(\tau^{(k)}, \tau_c^{(k)}, \theta) sign(\tau^{(k)}), \\
\dot{\gamma}^{(k)} = (\rho_+^{(k)} V_+^{'(k)} - \rho_-^{(k)} V_-^{'(k)}) |\mathbf{b}|^{(k)}, \\
\rho^{(k)} = f_2(\tau^{(k)}, \tau_c^{(k)}, \theta, \rho^{(k)}, \rho_{bar}^{(k)}), \\
\rho_{bar}^{(kl)} = f_3(\tau^{(k)}, \tau_c^{(k)}, \theta, \rho^{(k)}, \rho_{bar}^{(k)}), \\
\tau_c^{(k)} = f_4(\rho, \rho_{bar},).
\end{cases}
\tag{3}
$$

The average velocities of dislocations $(V_+^{'(k)}, V_-^{'(k)})$ play an important role for the description of the dislocation reactions, the most important role is played by the average velocities of the dislocations. In this case the velocities of motion of the dislocations of opposite signs may differ in sign and magnitude.

To determine the shear rates along the SS at the mesoscale-1, the rate form of the Orowan equation is used, including the densities of the positive and negative dislocations, the average slip velocities of dislocations (determined in the laboratory coordinate system (LCS)) and the modulus of the Burgers vector of dislocations:

$$
\dot{\gamma}^{(k)} = (\rho_+^{(k)} V_+^{'(k)} - \rho_-^{(k)} V_-^{'(k)}) |\mathbf{b}|^{(k)}.
\tag{4}
$$

The average velocities of motion of the dislocations are expressed by the following relation:

$$\begin{cases} V_+^{'(k)} = l^k v \, \exp(-\Delta G_*^k/k_B\theta) sign(\tau^{(k)}) \\ V_-^{'(k)} = -l^k v \, \exp(-\Delta G_*^k/k_B\theta) sign(\tau^{(k)}). \end{cases} \qquad (5)$$

The sign of the first factor in determining the average shear rates depends on the sign of the dislocations for which they are determined, e.g. for positive dislocations, the default sign will be a plus, for negative dislocations, it will be a minus.

3.1 Work of Dislocation Sources

As a rule, during plastic deformation, there is an increase in the density of dislocations on the SS. The Frank-Read sources that generate closed expanding dislocation loops are distinguished as intragranular dislocation sources. It is known experimentally that a source can generate a limited number of loops. The densities of the Frank-Read sources are introduced in the work. Until the critical stresses of the source are reached, no new loops are generated:

$$\begin{aligned} \rho_{0src}^{(k)} &= \rho_{0src}, \\ \dot\rho_{src}^{(k)} &= L\rho_{bar}^{kl}\rho^l[l^k v \, \exp(-\Delta G_*^k/k_B\theta)], \\ \tau_{src} &= \frac{A\mu b}{2\pi L}\left(\ln\frac{L}{r_0} + B\right), \end{aligned} \qquad (6)$$

where μ is the shear modulus, R_s is the average width of the fixed area (depends on the density of defects), b is the value of the Burgers vector, a is a dimensionless parameter. The contribution to the increase of the dislocation densities is proportional to the current density of the dislocation loops; and it is nonzero only when the acting stresses exceed the critical ones:

$$\dot\rho_{nuc}^{(k)} = k\rho_{src}H(\tau^{(k)} - \tau_{src}^{(k)}). \qquad (7)$$

3.2 Annihilation of Dislocations

Occurs as a result of a reaction on two dislocations with one SS and different extraplanes (in terms of the model, these are positive and negative dislocations).

This reaction is possible if two dislocations of the opposite signs are at a small distance from each other. The dislocation annihilation is most often observed in the experiments with reverse loading. The dislocations of the opposite signs in the same slip system attract. When the dislocations are located on the parallel slip systems, they can also creep towards each other, thus annihilating when approaching by the annihilation distance h_{ann}. The number of the reacted dislocations is proportional to the swept volume and dislocation density on the slip systems. To describe annihilation in the work, it is proposed to use the following relation:

$$\dot\rho_{\pm}^{(k)ann} = -h_{ann}\rho_+^{(k)}\rho_-^{(k)}\left|V_+^{'(k)} - V_-^{'(k)}\right|. \qquad (8)$$

3.3 Impurity Atoms

In the mesoscale-2 model, in addition to the interaction of the dislocations with each other, the interaction of the dislocations with the impurity atoms is introduced. The total amount of the impurity atoms is constant and can be distributed between the dislocations on the slip systems. We consider the interaction of the dislocations with the impurity atoms precipitated during the motion of the latter and being at the current time in the capture tube, which slow down the motion of the dislocations. Depending on the velocities of the dislocation motion, the concentration of the impurity atoms changes in the volume of the capture tube, so the contribution to the critical shear stresses along the SS from the action of the impurity atom changes. The change in the concentration of the impurity atoms captured by the dislocations on the k-th slip system is determined according to the two scenarios.

The First Scenario. If the average velocity of dislocations $V^{(k)}$ is lower than the impurity atoms velocity caused by self-diffusion $V_{diff.}$ at a given temperature, and the current concentration of the impurity atoms $c_{capt}^{(k)}$ captured by the dislocation is lower than the critical concentration $c_{crit}^{(k)}$ (the highest concentration of the atoms that the dislocation can hold), in this case the dislocation collects the impurity atoms in its capture tube. A new concentration of the captured and free impurity atoms is calculated depending on the value of the dislocation velocity. In this case, the greater the dislocation velocity, the more impurity atoms the dislocation collects per unit time. However the concentration of the captured atoms cannot exceed the critical one. The change in the concentration of the captured atoms in this case can be expressed by the following relation:

$$\dot{c}_{capt1.1}^{(k)} = c_{free}^{(k)} \frac{V^{(k)}}{R} H\left(V_{diff.} - \left|V^{(k)}\right|\right) H\left(c_{crit.}^{(k)} - \varepsilon - c_{capt.}^{(k)}\right), \tag{9}$$

where ε is a small positive value, for example, 0.1% of the critical concentration.

If the average velocity of the dislocation motion $V^{(k)}$ is lower than the impurity atoms velocity caused by self-diffusion $V_{diff.}$ and the current concentration of the impurity atoms $c_{capt.}^{(k)}$ captured by the dislocation equals to the critical concentration $c_{crit.}^{(k)}$, in this case the dislocation moves together with the cloud of the impurity atoms of a constant (equal to the limit one) concentration. At this time, the cloud is simultaneously resupplied with new captured impurity atoms, and it releases excessive atoms, then both processes are continuous. The concentration of the captured and free impurity atoms remains the same, so the rate of changing the number of the captured atoms should become zero, which is described by relation (8). As soon as the concentration of the captured atoms becomes critical (taking into account the introduced tolerance), due to the second Heaviside function, the source capture is set to zero.

The Second Scenario. If the average velocity of the dislocation $V^{(k)}$ is higher than the impurity atoms velocity caused by self-diffusion $V_{diff.}$, then there is a new calculation of the concentration of the captured and free impurity atoms depending on the value of the dislocation velocity. The higher the dislocation velocity, the more impurity atoms throw off the dislocation per unit time. When moving, the front wall of the capture tube sweeps a certain volume per unit time, in which there are impurity atoms with the concentration of free atoms. In other words, it determines the intensity of the source of absorption by the capture tube per unit time. The back wall also sweeps a certain volume, in which the impurity atoms remain with a concentration equal to the concentration of the captured atoms at the current time. This determines the intensity of the sink of the impurity atoms from the capture tube (the amount is equal to the product of the concentration of the captured atoms by the noticeable volume) and is described by the relation in (9):

$$\dot{c}_{capt2} = \left(c_{free} - c_{capt} \right) \frac{V^{(k)}}{R} H \left(\left| V^{(k)} \right| - V_{diff.} \right). \tag{10}$$

The general system of equations determining the contribution from the interaction of the dislocations with the impurity atoms is as follows:

$$\begin{cases} \dot{c}_{capt.}^{(k)} = c_{free.}^{(k)} \dfrac{V^{(k)}}{R} H \left(c_{\kappa crit.}^{(k)} \text{-}\varepsilon - c_{capt.}^{(k)} \right) H \left(V_{diff.} - \left| V^{(k)} \right| \right) + \\ \qquad + \left(c_{free} - c_{capt} \right) \dfrac{V^{(k)}}{R} H \left(\left| V^{(k)} \right| - V_{diff.} \right), \\ c^{(k)} = c_{capt.}^{(k)} + c_{free.}^{(k)} = \text{const}, \\ \dot{\tau}_{imp}^{(k)} = \alpha \tau' \dfrac{\dot{c}_{capt.}^{(k)}}{c^{(k)}}. \end{cases} \tag{11}$$

3.4 Hardening Law

A hypothesis is accepted about the possibility of an additive decomposition of the critical stresses of the SS into contributions from the lattice resistance (constant value), from the stress fields of the dislocations and impurity atoms:

$$\begin{aligned} \tau_{c0}^{(k)} &= \tau_{lat}^{(k)}, \\ \dot{\tau}_{c}^{(k)} &= \dot{\tau}_{disl}^{(k)} + \dot{\tau}_{imp}^{(k)}, \\ \dot{\tau}_{disl}^{(k)} &= \alpha \sum_{i=1}^{n} M^{ik} \dot{\rho}^i, \\ \dot{\tau}_{imp}^{(k)} &= \alpha \tau' \frac{\dot{c}_{capt.}^{(k)}}{c^{(k)}}. \end{aligned} \tag{12}$$

4   Model Implementation Algorithm

Computations based on the model algorithm at all scale levels are implemented in terms of physical time. The stability of the numerical solution depends on the choice of the time step. The considered model can be integrated into the general model for solving boundary value problems of loading structures. In this case the element of the macrolevel will describe the mechanical response of the element of the considered finite element mesh. An algorithm for the implementation of one step (in time) of the deformation process will be described below. It should be noted that the plastic deformation does not usually occur immediately. To begin the plastic deformation, it is necessary to fulfill a certain criterion, the first stages of deformation usually occur in an elastic manner. Such staging imposes restrictions on the size of the time step, and may also require its significant reduction at the time of changing modes. The model is formulated in a velocity form, for integration over time a step-by-step process and integration by the Euler method are used.

At the macrolevel, the deformation of the representative macrovolume of the material (macrodot) is considered. The macrolevel parameters include components of the Cauchy stress tensor (current value and rate of change at the current time step) and components of the spin tensor. Structurally, the macrolevel consists of a set of mesoscale-1 elements, while each element has its own characteristics, such as the orientation of the lattice in space, stress, etc. Deformations are prescribed to the macrodot, by using the Voigt hypothesis, the deformations are transferred to the mesoscale-1. In each element of the meso-1, stresses, the rate of rotation of the lattice, and the magnitude of plastic deformations are determined. By averaging the characteristics of the meso-1 elements (the averaging assumes the equality of the representatives of the meso-1 included in the macrolevel, i.e., the averaging over the representative macrovolume takes place), the stresses at the macrolevel are determined (Fig. 1).

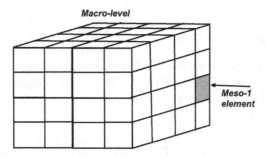

Fig. 1. A schematic representation of a macrolevel element consisting of a set of the meso-1 elements.

At the meso-1, the mechanical behavior of the crystallite is described, that is a region with a uniform crystal lattice. The mesoscale-1 parameters include the components of the Cauchy stress tensor (at the current step and in velocities), shear rates at the SS of the current step, the components of the spin tensor at the current step, and the values of shear stresses at each SS. At the mesoscale-1, it is assumed that the mesostresses, the rate of rotation of the lattice, and the rates of shear along the SS are homogeneous. The input variables are the components of the strain rate tensor, the response is the Cauchy stress tensor and the rate of the lattice rotation. At the mesoscale-1, the characteristics of the stress-strain state of the crystallite are determined, the magnitude of inelastic deformations, using Hooke's law in the rate relaxation form, the stress changing rate and stress at the current step are determined. The value of the stress tensor (calculated in the previous step) is used to determine the magnitude of the shear stresses that are subsequently transmitted to the meso-2 level. From the mesoscale-2 to the mesoscale-1, shear rates are transmitted along the SS, determined from the velocities of motion and dislocation densities (Fig. 2).

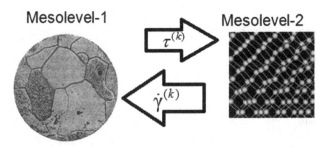

Fig. 2. A schematic representation of the relation between elements of the meso-1 and meso-2.

At the mesoscale-2, the evolution of densities and the kinetics of structural defects (edge dislocations and impurity atoms) are described. The mesoscale-2 parameters include the dislocation densities along the SS (in velocities and absolute values), critical stresses at each SS (at velocities and absolute values), average slip velocities of the dislocations at each SS, temperature, the concentrations of the free atoms and impurities. From the shear stresses transferred from the meso-1 and critical stresses from the previous step, we determined the average slip velocities of dislocations, the number of dislocation sources formed, the rate of change in the density of dislocations emitted by the sources, the rate of change in density due to the annihilation of the dislocations, the number of fixed and free impurity atoms. The rates and the absolute value of critical stresses for the next integration step are determined from the rates of change in the dislocation densities and the concentrations of the captured impurities. The absolute dislocation densities along the SS are found, using the Orowan equation, in the velocities, the shear rates transmitted to the meso-1 are determined.

In accordance with the structure of the model, the algorithm for its implementation is carried out by a stepwise (in physical time) solution of the following set of equations:

Macrolevel:

input : $\mathbf{Z}, - >$Meso1;

Meso1 : |

$\mathbf{z} = \mathbf{Z}$

$\tau^{(k)} = \mathbf{b}^{(k)}\mathbf{n}^{(k)} : \sigma, - >$ Meso2;

Meso2 :

$$v_{\pm}^{'(k)} = \pm f\left(\tau^{(k)}, \tau_c^{(k)}, \theta\right)\text{sign}\left(\tau^{(k)}\right),$$

$$\dot{\rho}_{\pm}^{(k)} = \dot{\rho}_{muc\pm}^{(k)} + \dot{\rho}_{ann}^{(k)},$$

$$\dot{\rho}_{muc}^{(k)} = f\left(\rho_s^{(k)}, \tau^{(k)}, \theta\right),$$

$$\dot{\rho}_{\pm}^{(k)ann} = -h_{ann}\rho_+^{(k)}\rho_-^{(k)}\left|v_+^{'(k)} - v_-^{'(k)}\right|$$

$$\dot{c}_{capt.}^{(k)} = c_{free.}^{(k)}\frac{V^{(k)}}{R}H\left(c_{crit.}^{(k)} - \varepsilon - c_{capt.}^{(k)}\right)H\left(V_{diff.} - |V^{(k)}|\right) + \left(c_{free} - c_{capt}\right)\frac{V^{(k)}}{R}H\left(|V^{(k)}| - V_{diff.}\right),$$

$$\dot{t}_{disl}^{(k)} = \alpha\sum_{i=1}^{n}M^{ik}\dot{\rho}^i,$$

$$\dot{t}_{imp}^{(k)} = \alpha\tau'\frac{\dot{c}_{cap.}^{(k)}}{c^{(k)}},$$

$$\dot{t}_c^{(k)} = \dot{t}_{dis}^{(k)} + \dot{t}_{imp}^{(k)},$$

$$\dot{\gamma}^{(k)} = \left(\rho_+^{(k)}v_+^{'(k)} - \rho_-^{(k)}v_-^{'(k)}\right)|\mathbf{b}|^{(k)}, - >$$ Meso1

Meso1:

$$\mathbf{z}^{in} = \sum_{k=1}^{K}\dot{\gamma}^{(k)}\mathbf{b}^{(k)}\mathbf{n}^{(k)},$$

$$\omega = \dot{\mathbf{o}}\cdot\mathbf{o}^T,$$

$$\mathbf{z} = \widehat{\nabla}\mathbf{v}^T - \boldsymbol{\omega},$$

$$\sigma^{cr} = \Pi : \left(\mathbf{z} - \mathbf{z}^{in}\right), - >$$ Macrolevel

$$(13)$$

It should be noted that the elements of the meso-1 and meso-2 in this formulation can be implemented simultaneously for one macrovolume, which makes it possible to implement the parallel algorithm for these elements.

5   Determining Model Parameter Values

Benefits of the physically-oriented models include the possibility of determining the values of parameters from numerical experiments on simplified research objects. Often, such work is carried out using mathematical packages implemented in functional programming languages, for example Wolfram Mathematica, Matlab, etc. The disadvantages of these tools include an inconvenience for implementing large algorithms, the complexity of implementing modular programs. To solve small applied problems, the mathematical packages offer great advantages; they have great function libraries, offer an easy visualization of results, as well as all the advantages of the high-level paradigm of the package implementation.

Additional numerical processes were used to determine some of the parameters. When determining the critical stresses, which depend on the interaction of the edge dislocations with each other, the solution is used for an isotropic elastic medium with a fixed single dislocation:

$$
\begin{cases}
\sigma_{xx} = -L \cdot \dfrac{y(3x^2 + y^2)}{(x^2 + y^2)^2}; \\[2mm]
\sigma_{yy} = -L \cdot \dfrac{y(x^2 - y^2)}{(x^2 + y^2)^2}; \\[2mm]
\sigma_{xy} = -L \cdot \dfrac{x(3x^2 - y^2)}{(x^2 + y^2)^2}; \\[2mm]
\sigma_{zz} = \nu(\sigma_{xx} + \sigma_{yy}); \; \sigma_{xz} = \sigma_{yz} = 0.
\end{cases}
\tag{14}
$$

The dislocations from another SS were introduced into the neighborhood of the fixed dislocation segment, and the forces of interaction of such segments were estimated. The forces with which two dislocations interact were considered (in the case of dislocations, the dislocation segments were taken from different SSs), and the possible variants of their interposition at different distances were taken into account. By averaging these forces, we could estimate the interaction forces between the dislocations on the fixed SS and introduced dislocations, and to construct a matrix describing the interaction intensity of the dislocations with different SSs. When determining the components of this matrix, the Wolfram Mathematica application package was used.

The stresses generated by the deposition of the impurity atoms on the edge dislocations were determined in a similar way, and the characteristics of the process of capturing the impurity atoms by the edge dislocations were determined.

6 Software Implementation, Results

The above model was implemented in C ++ in a procedure-oriented paradigm. The results of applying the algorithm to describe the evolution of the defects density in a crystallite (the meso-1 and meso-2 systems) and a polycrystalline aggregate of 64 elements (the macrolevel, the meso-1 and meso-2) are presented. To describe all the main processes, separate functions were used; for all real values, the double-precision double type was used. The results visualization is done in the packages Wolfram Mathematica and VeusZ.

The implementation of the parallel computations was made using the Intel c ++ compiler, the elements of the meso-1 and meso-2, as well as a number of procedures, were subject to parallel execution. The developed algorithm can be built into the finite element software products, which allows us to solve applied problems to describe the response of structures.

In the numerical implementation, a time step of 0.001 s^{-1} was used. The elasto-plastic deformation of the AMg6 alloy was simulated, and the strain rates for shear ($z_{12} = z_{21} = 1 \cdot 10^{-4} \text{s}^{-1}$) were set as the loading. In the reference configuration, the crystallographic coordinate system was assumed to coincide with the laboratory one. Some of the parameters were taken from references; others were determined by an additional research. The deformation was performed until the intensity of the accumulated deformation reached 10%.

The numerical implementation of the model allows one to analyze the magnitude of all the variables during the deformation, which makes it possible to analyze the evolution mechanisms of the lattice defects (Fig. 3).

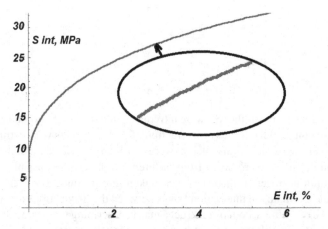

Fig. 3. The dependence of stresses on deformations (a single crystal, experiment 1).

During deformation, the main attention was paid to the fixation and release of the edge dislocations from the impurity atoms. The existing model makes it possible to analyze all the parameters of the system during the plastic deformation, as well as to analyze the influence of all the evolution mechanisms of the defect densities. After a hundred of iterations, the main characteristics of the system were recorded in a file, which made it possible to further analyze them, as well as debug the program code. After the onset of the plastic deformation, an increase in the dislocation densities and an active redistribution of the concentration of the impurity atoms on the active SS were observed in both experiments (Fig. 4).

In the experiments with polycrystals, a higher level of stress jumps was observed, which is explained by the gradual accumulation of the impurity atoms on the dislocations in many crystallites, and then by their release in an avalanche-like manner (Fig. 5).

When analyzing the processes of pinning and releasing the impurity atoms, the dependence of the concentration of the impurity atoms fixed on the SS on the accumulated deformation was considered. Let us note that this process is more active at the initial stage of the plastic deformation. Qualitatively, this phenomenon is observed under plastic loading in field experiments.

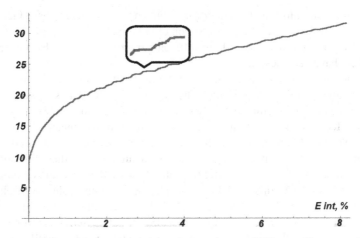

Fig. 4. A dependence of stresses on deformations (polycrystal, 64 crystallites, experiment 2).

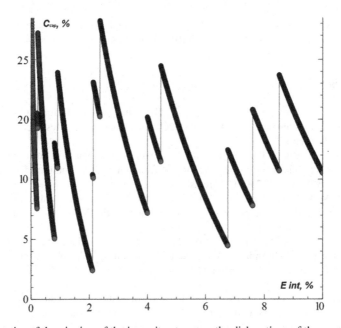

Fig. 5. Kinetics of the pinning of the impurity atoms on the dislocations of the considered SS.

7 Conclusion

This article gives an overview of the models describing the Portevin - Le Chatelier effect and outlines the main advantages and disadvantages of the existing models. An attempt has been made to use a common approach including the multilevel models based on crystal plasticity to describe the discontinuous plasticity effect. The model previously developed by the team of authors was modified taking into account the presence of the impurity atoms in the alloy and their interaction with the edge dislocations, which affects the critical slip stresses. A detailed description of the direct elastoviscoplastic model is given including the relations describing different structural levels. Also the relations between them have been described.

The model implementation algorithm for the material representative volume at all scale levels is given. This model is a part of the software package that is implemented and developed in a modular structure that allows one to describe a wide range of physical and mechanical effects. To determine the values of a number of parameters, the additional package of Wolfram Mathematica for the mathematical calculations was used.

We identified the model parameters, carried out a number of studies and showed the qualitative agreement of the results with the data of the field experiments. Some of the computation results using this model are considered, the main evolution mechanisms of the microstructure are described. The calculations were made using an algorithm implemented in the C++ language in the VisualStudio 2012 series environment; the Intel Parallel Studio development package was used to implement the parallel computing procedures.

Acknowledgments. The work was supported by the Russian Ministry of Education and Science (the basic part of the State Assignment for PNRPU, Project No. FSNM - 2020-0027).

References

1. Deryugin, E.E., Panin, V.E., Schmauder, Z., Storozhenko, I.V.: Deformation localization effects in composites based on Al with Al2O3 inclusions. Phys. Mesomech. 3(4), 35–47 (2001). (in Russian)
2. Deryugin, E.E., Panin, V.E., Schmauder, Z., Suvorov, B.I.: Study of local characteristics of intermittent fluidity of dispersion-hardened aluminum as a multilevel system. Phys. Mesomech. 5(9), 27–32 (2006). (in Russian)
3. Lebedkin, M.A., Dunin-Barkovsky, L.R., Lebedkina, T.A.: Statistical and multifractal analysis of collective dislocation processes under conditions of the Portevin – Le Chatelier effect. Phys. Mesomech. 2(4), 13–19 (2001). (in Russian)
4. Shibkov, A.A., Denisov, A.A., Zheltov, M.A., Zolotov, A.E., Hasanov, M.F., Kochegarov, S.S.: Suppression of intermittent deformation of Portevin – Le Chatelier by constant electric current in the aluminum-magnesium alloy AMg5. Solid State Phys. 57, 228–236 (2015). (in Russian)
5. Shibkov, A.A., Zolotov, A.E.: Nonlinear dynamics of spatiotemporal structures of macrolocalized deformation. JETP Lett. 5(90), 412–417 (2009). (in Russian)
6. Trusov, P.V., Chechulina, E.A.: Discontinuous fluidity: physical mechanisms, experimental data, macrophenomenological models. Bull. Perm Nat. Res. Polytech. Univ. Mech. 3, 186–232 (2014). https://doi.org/10.15593/perm.mech/2014.3.10
7. Estrin, Y., Kubin, L.P: Local strain hardening and nonuniformity of plastic deformation. Acta Metall. 34, 2455–2464 (1986)
8. Chen, Z., Zhang, Q., Wu, X.: Multiscale analysis and numerical modeling of the Portevin – Le Chatelier effect. Int. J. Multiscale Comput. Eng. 2(3), 227–237 (2005). https://doi.org/10.1615/intjmultcompeng.v3.i2.70
9. Hopperstad, O.S., Børvik, T., Berstad, T., Lademo, O.-G., Benallal, A.: A numerical study on the influence of the Portevin–Le Chatelier effect on necking in an aluminium alloy. Model. Simul. Mater. Sci. Eng. 15, 747–772 (2007). https://doi.org/10.1088/0965-0393/15/7/004

10. McCormick, P.G.: Theory of flow localization due to dynamic strain ageing. Acta Metall. **12**(36), 3061–3067 (1988). https://doi.org/10.1016/0001-6160(88)90043-0
11. Kok, S., et al.: Spatial coupling in jerky flow using polycrystal plasticity. Acta Mater. **51**, 3651–3662 (2003). https://doi.org/10.1016/s1359-6454(03)00114-9
12. Keshavarz, S., Ghosh, S.: Hierarchical crystal plasticity FE model for nickel-based superalloys: sub-grain microstructures to polycrystalline aggregates. Int. J. Solids Struct. **55**, 17–31 (2015)
13. Smirnov, A.S., Konovalov, A.V., Muizemnek, O.Y.: Modelling and simulation of strain resistance of alloys taking into account barrier effects. Diagn. Resour. Mech. Mater. Struct. **1**, 61–72 (2015). (in Russian)
14. Trusov, P.V., Shveykin, A.I.: Multilevel models of mono- and polycrystalline materials: theory, algorithms, application examples. SO RAN, Novosibirsk (2019). https://doi.org/10.15372/multilevel2019tpv. (in Russian)
15. Trusov, P.V., Chechulina, E.A.: Discontinuous fluidity: models based on physical theories of plasticity. Bull. Perm Nat. Res. Polytech. Univ. Mech. **1**, 134–163 (2017). https://doi.org/10.15593/perm.mech/2017.1.09
16. Mesarovic, S.D.: Dynamic strain aging and plastic instabilities. J. Mech. Phys. Solids **5**(43), 671–700 (1995)
17. Cottrell, A.H.: Dislocations and plastic flow in crystals. GNTI Literature on Ferrous and Non-Ferrous Metallurgy, Moscow (1958). (in Russian)
18. Ham, F.S.: Stress-assisted precipitation on dislocations. J. Appl. Phys. **6**(30), 915–926 (1959). https://doi.org/10.1063/1.1735262
19. Bullough, R., Newman, R.C.: The kinetics of migration of point defects to dislocations. Rep. Prog. Phys. **1**(33), 101–148 (1970). https://doi.org/10.1088/0034-4885/33/1/303
20. Balik, J., Lukáč, P.: Influence of solute mobility on dislocation motion. I. Basic model. Czech. J. Phys. **39**, 447–457 (1989)
21. Balik, J., Lukáč, P.: Influence of solute mobility on dislocation motion. II. Application of the basic model. Czech. J. Phys. **39**, 1138–1146 (1989)
22. Aboulfadl, H., Deges, J., Choi, P., Raabe, D.: Dynamic strain aging studied at the atomic scale. Acta Mater. **86**, 34–42 (2015)

Mathematical Software for Testing and Setting up the Induction Soldering Process

Anton Milov[1]([⊠]) [iD], Vadim Tynchenko[1,2,3] [iD], Vyacheslav Petrenko[1] [iD], and Vladislav Kukartsev[1,2] [iD]

[1] Reshetnev Siberian State University of Science and Technology, Krasnoyarsky Rabochy Avenue 31, 660037 Krasnoyarsk, Russia
antnraven@ieee.org, vadimond@mail.ru
[2] Siberian Federal University, Svobodny Avenue 79, 660041 Krasnoyarsk, Russia
[3] Marine Hydrophysical Institute of Russian Academy of Sciences, капитанская Street 2, 299011 Sevastopol, Russia

Abstract. This paper presents mathematical software designed to test and adjust the technological process of induction soldering of waveguide paths. The development of the technological process in the form of full-scale experiments is an expensive and time-consuming process. The use of mathematical software will reduce the labor intensity and cost of development of the technological process due to the fact that the development of the field experiments will be made on the basis of technological parameters selected on the basis of model data. The mathematical software is a complex of mathematical models of both individual elements and the entire assembly of the waveguide path as a whole. The software is the implementation of these mathematical models to test and configure the induction soldering process. Visual Studio Community 2019 is used as a development tool. The license conditions allow using this tool for academic research purposes. The software is developed within the framework of the object-oriented approach. Using this software will simplify and reduce the cost of testing and setting up the technological process of induction soldering of waveguide paths.

Keywords: Induction soldering · Software · Technological process · C# · Visual studio · Mathematical model

1 Introduction

Features of the technological process of creating permanent joints on the basis of induction heating is that the initial setting of the process technological parameters has the most influence on the process quality. Most often, the optimal mode of the induction soldering technological process is determined experimentally using control samples. This approach to determining the optimal mode is quite laborious and expensive.

The simplest way to increase the efficiency of the process of induction soldering of waveguide paths is to develop mathematical models and software designed to test this technological process, which will significantly reduce costs.

V. Jordan et al. (Eds.): HPCST 2020, CCIS 1304, pp. 114–124, 2020.
https://doi.org/10.1007/978-3-030-66895-2_8

The use of the proposed approach within the framework of this work will simplify and reduce the cost of testing and adjusting the induction soldering process, which will allow quickly readjusting the technological process to different standard sizes of the same type of products and start-up of the production process of other types of products.

2 Literature Review

The work [1] presents the development of a multiphysics model of the technological process for the production of modular solar panel systems. Many branches of mechanical engineering [3–5], as well as the production of solar panels, waveguide paths [2], are based on induction soldering. The use of models allows you to improve the quality of the technological process, which in turn has a positive effect on the quality of products, both in the photovoltaic and aerospace industries. The model developed in the framework of research [6] focuses on the behavior of the solder during heating and melting. Based on the simulation result, a technique is proposed, in which the number of voids at the joints is significantly reduced. Chinese researchers [7] have created models of heat distribution in aluminum sheets. The authors of this work have identified the most significant factors affecting the quality of the technological process of hot stamping of aluminum sheets using induction heating. An analytical model for calculating the parameters of inductance coils was proposed in [8]; the results of modeling by the authors will be used in the future to work out the process of soldering aluminum blanks.

The authors of [9] presented a mathematical model of the technological process based on induction heating using the commercial package Cedrat Flux 10.3. The reliability of the proposed model has been confirmed in the framework of experimental studies. The simulation results are in good agreement with the experimental temperature profile at specified points on the surface of the part. The advantage of the mathematical model proposed by the authors is the possibility of using the model to predict such parameters as current density and magnetic flux field inside the workpiece, which are difficult to directly measure.

Research [10] shows that decreasing the order of models is a rather effective and promising tool for controlling the technological processes of induction soldering (TPIS) based on indirect measurements. Based on Propper's orthogonal decomposition method, a fourth-order system was obtained. The resulting model ensures the achievement of the target temperature, while its computational complexity is low enough for its prompt solution using a fairly inexpensive microcontroller. Within the framework of research [10], a thermal model of the technological process of infrared reflow soldering was developed. The model predicts the associated thermal effects, from the convection characteristics in an infrared oven to a detailed thermal response, including the change in the transition of the solder between the solid and liquid phases.

A neural network model for controlling the TPIS is presented in [11]. The use of the model proposed by the authors makes it possible to significantly improve the quality of control of the technological process of induction soldering of waveguide paths (TPIS WP).

The study [12] demonstrates a mathematical model for controlling the TPIS of waveguide paths in the form of a neuro-fuzzy controller. The effectiveness of the proposed model has been confirmed in the framework of field experiments. The use of simulation modeling for modeling the TPIS is demonstrated in [13, 14]. Such widespread use of simulation modeling for the optimization of technological processes based on induction heating for various branches of mechanical engineering makes it possible to speak about the high efficiency of this approach to modeling of the technological processes under consideration.

Improving the quality of control of the TPIS of waveguide paths is possible by introducing additional control loops and using intelligent methods [15]. However, at the same time, the introduction of an additional circuit does not completely solve the problems with uncertainty conditions in which the course of the technological process takes place. And a lot of experimental data is required for teaching intelligent methods, the receipt of which entails serious costs for conducting experiments on control samples.

3   Mathematical Support

As a software for testing and tuning the TPIS, a complex of mathematical models has been developed individually for the elements of the waveguide path assembly and for the assembly of the waveguide path as a whole. Structurally, the waveguide path of the spacecraft consists of connectable pipes and a flange /coupling for their connection.

An instant heating source in a flat rod is used as a mathematical model for heating assembly elements in order to work out the TPIS WP:

$$T(x, t) = \int_0^t \frac{Q}{c\rho F \sqrt{4\pi at}} e^{(-\frac{x^2}{4at} - bt)} \tag{1}$$

$$b = \frac{\alpha p}{c\rho F} \tag{2}$$

where Q – heat amount [J], F – pipe cross-section [m^2], x – distance from heat source [m], ρ – volumetric heat capacity [J/m^3], t – time [sec], b – thermal coefficient convection into the external environment from the surface of the rod, formula (2), a – thermal conductivity coefficient, p – perimeter of the section.

$$T(x, t) = \int_{i=-\infty}^{\infty} \int_{j=-1,1} \frac{Q}{c\rho F \sqrt{4\pi at}} e^{(-\frac{(x-jl-2iL)^2}{4at} - bt)} \tag{3}$$

where L – rod length [m], l – distance from the end to the heating source [m].

Calculation formula (4) for heating a waveguide tube with reference to a specific standard size in this case is:

$$T(x, t) = \sum_{j=-1,1} \frac{Q}{Fc\rho \sqrt{4at}} e^{(-\frac{(x+jl)^2}{4at} - bt)}. \tag{4}$$

Calculation formula (5) for heating the flange of the waveguide assembly with reference to a specific standard size has the form:

$$T(x, t) = \sum_{i=0}^{j} \frac{2Q}{Fc\rho\sqrt{4at}} e^{\left(-\frac{(x+2iL)^2}{4at} - bt\right)} \tag{5}$$

where j – number of reflections taken into account in the calculation, selected in such a way that for $j + 1$ for any x and t: $T(x,t) \leq \varepsilon$ at $\varepsilon \to 0$.

The fundamental difference between the heating modes of pipe-flange-pipe, pipe-sleeve-pipe, and pipe-flange, pipe-sleeve assemblies is in the law of distribution of heating energy between the corresponding elements of the wave-water path assembly. When modeling the control system, we take it as an assumption that there is a certain law of energy distribution between the elements of the waveguide assembly, which is tied to both the standard size and the configuration of the inductor. The technology of induction soldering of waveguide paths involves the use of inductors with a beveled window, which ultimately makes it possible to localize the zones of maximum heating near the soldering zone.

Based on this feature of the technology, we can assume in modeling that all the energy transferred by the inductor is released in the soldering zone. Thus, for the convenience of modeling, it is assumed that all the generator energy will be transferred to the soldered assembly. Thus, we formulate the law of distribution of the amount of heat between the elements of the wave-water assembly in the following form:

$$q(t) = q(t)M(x) + q(t)(1 - M(x)) \tag{6}$$

where $q(t)$ – continuously operating heat source; $M(x)$ – coefficient of heat distribution between the elements of the assembly. $M(x)$ belongs to [0,1] and x belongs to $[n,m]$, where n and m are the lower and upper boundaries, respectively; x is the distance from the flange/coupling to the inductor window.

Distribution function (6) was obtained by the authors empirically.

4 Software Support

The mathematical models presented in the previous paragraph for testing and tuning the TPIS must be implemented as a prototype of a software product.

The software product is a classic Windows Forms application [16], implemented in the C # programming language [17–19] with the object-oriented programming paradigm [20]. The integrated development environment is Visual Studio 2019 Community [21], whose use for academic research is permitted under the license.

The purpose of the software product is to use it for testing and adjusting the TPIS WP. The module allows you to set such parameters of the induction soldering technological process as: the stabilization temperature of the technological process, the power of the induction heating source, the geometric dimensions of the assembly elements of the waveguide path. The mathematical model created on the basis of the given parameters can be used to test the TPIS. In addition, the functionality of the software product allows you to choose a process control algorithm, including the choice of both classical control methods and intelligent ones.

The structure of the software module for testing and tuning the TPIS WP is shown in Fig. 1.

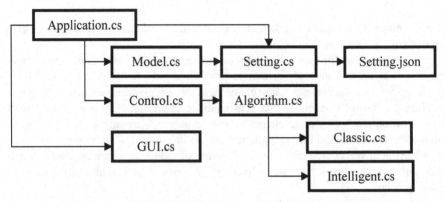

Fig. 1. Software structure scheme.

The structure of the software product, shown in Fig. 1, fully corresponds to the composition of the classes, on the basis of which the algorithms for the operation of the application are implemented. The most convenient way of description in this case is the class diagram (Fig. 2) in UML notation [22, 23, 24], the description of which will be presented below.

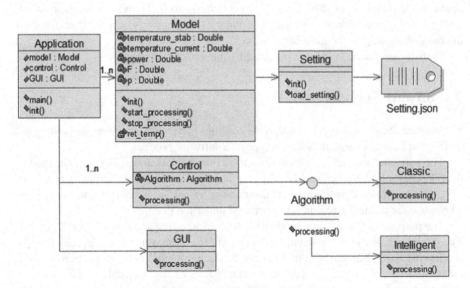

Fig. 2. Software class diagram.

From the class diagram, you can see that the Application class is responsible for the operation of the main application algorithms. In this class, calls of all classes are implemented, which solve specific problems necessary for working out the TPIS. In particular, the execution of the application includes the initialization of all the necessary objects, such as the mathematical model itself, the simulated control algorithm, elements of the graphical user interface, as well as handlers for interaction with these elements.

The Model class implements the very mathematical model of heating the wave-water duct assembly for working out the induction soldering process. The class is responsible for setting the necessary settings of the mathematical model, such as: stabilization temperature, power of the induction heating source, geometric dimensions of the assembly elements of the waveguide path of the spacecraft. The class is also responsible for starting and stopping the induction soldering process simulation.

The Setting class, which is responsible for loading model parameters, is closely related to this class. Loading is possible from the Setting.json configuration file, and it is possible to use the data entered by the user into the elements of the graphical interface.

The Control class implements the control algorithm used in the development of the TPIS WP. The class itself is only needed to call the algorithms implemented in the Classic and Intelligent classes. Algorithms from these classes are called through the Algorithm interface. The use of the interface makes it possible to uniformly use algorithms that implement both classical methods of controlling the TPIS WP and intelligent methods. This approach greatly simplifies the work with the program code and will also make it very easy and without artifacts to add new algorithms for controlling the technological process.

The GUI class is responsible for all the algorithms for interacting with the graphical user interface. The implementation details of all algorithms presented in the above classes, except for the Application and GUI classes, are encapsulated. All signatures are designed in such a way that they can be easily connected to other developed software modules as well. For example, it is intended for subsequent use in modules designed to test other similar technological processes for creating permanent joints.

Figure 3 shows a block diagram of the operation algorithm of the software module.

As you can see from the figure, the general progress of the application looks as follows. At the initial stage, the main application is initialized. At this stage, all application objects are created. Then you can choose from two options for loading settings. The settings can be obtained from the json configuration file, or obtained from the corresponding elements of the graphical user interface. In the next step, the settings are loaded into the mathematical model.

The next stage is the choice of the technological process control algorithm, on the basis of which the technological process will be worked out. Then the model is launched. The model is executed until the stabilization temperature is reached, after which it is possible to terminate the application or re-execute the model with the same or new settings.

Figure 4 shows the appearance of the application window. The application is one-window. As you can see in Fig. 4, the application window can be conditionally divided into four zones. The first zone contains the settings of the mathematical model, which include the stabilization temperature, the power of the induction heating source, and the

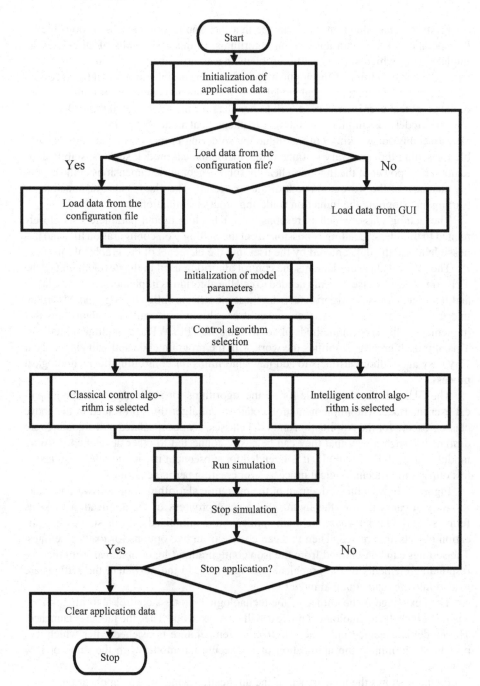

Fig. 3. Software block diagram.

standard size of the elements of the waveguide path assembly. In the second zone, there are settings for the control algorithm for the technological process of induction brazing of thin-walled aluminum waveguide paths of the spacecraft. In the third zone there is a graph of the TPIS WP. In the fourth zone there are buttons for starting and stopping the simulation of the TPIS.

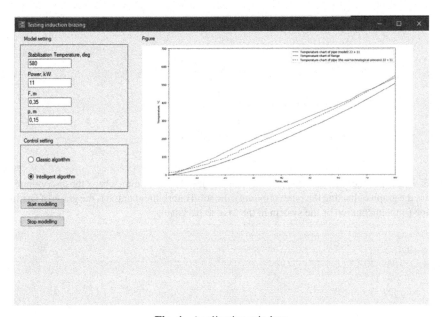

Fig. 4. Application window.

This software module is a part of a more general control system for the technological process of induction brazing of wave-water paths. Within the framework of the general interaction, the input data for the module are the target parameters of the technological process, as well as the type of the control method: classical or intelligent algorithms. The result of the module is a trained model that can be loaded into the main control system to set the optimal parameters of the technological process.

5 Experimental Verification

To check the efficiency of the developed software, we simulated the induction heating process for waveguide circuit assemblies with different standard sizes and compared it with data from real technological processes with the same standard sizes of waveguide circuit assembly elements.

Figures 5 and 6 show comparative graphs of a model of the TPIS of thin-walled aluminum waveguide paths of the spacecraft with graphs of real technological processes with the same standard sizes as those in the model data.

Temperature, °C

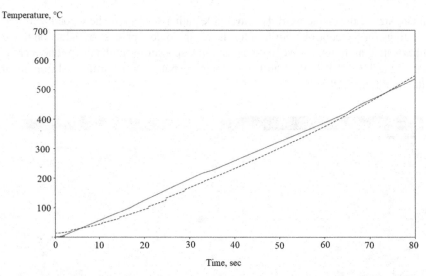

Time, sec

Fig. 5. The experiment graph for a pipe of a standard size of 22 × 11 mm, where: the discontinuous graph is the graph of heating the pipe (program); the solid intermittent graph is the graph of heating the pipe (implementation of the system in the Matlab package).

Temperature. °C

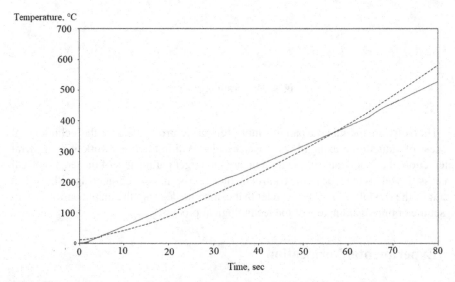

Time, sec

Fig. 6. The experiment graph for a pipe of a standard size of 19 × 9.5 mm, where: the discontinuous graph is the graph of heating the pipe (program); the solid intermittent graph is the graph of heating the pipe (implementation of the system in the Matlab package).

As the above graphs show, the developed model of the TPIS WP to a fairly high degree corresponds to the real flow of the process of induction soldering of waveguide paths of the spacecraft.

In all two cases, it can be seen that the results of modeling the heating of the elements of the waveguide duct assembly correspond to the real TPIS WP to a sufficiently high degree of accuracy.

6 Conclusion

Within the framework of this work, the mathematical and software support for the development and adjustment of induction brazing of thin-walled aluminum waveguide paths of the spacecraft is presented. The software is a set of mathematical models for heating individually the elements of the assembly of the waveguide path and the entire assembly as a whole. The use of the mathematical models presented in this work for testing the TPIS of thin-walled aluminum waveguide ducts will allow many different experiments to be performed to develop various soldering modes of various configurations of waveguide duct assemblies.

In addition, within the scope of this research, software has been developed and presented in the form of a Windows Forms application in the C # programming language. The use of this software will allow for the development of the TPIS based on the developed models, considering all the necessary parameters, as well as using both classical and intelligent control methods. The use of software will reduce the cost of research and development, because thanks to the model, full-scale tests can be carried out on already developed technological processes, which will reduce the cost of consumables.

Further research will be aimed at developing mathematical models for heating products in other technological processes in order to create non-detachable joints. Such processes can be electron beam welding and diffusion welding.

Acknowledgments. This research was financially supported by Reshetnev Siberian State University of Science and Technology project "Mathematical software of the induction soldering of spacecraft's waveguide paths".

References

1. Zeller, U., Lohmeier, M., Pander, M., Lausch, D.: Multiphysics simulation of induction soldering process. In: 2018 IEEE 7th World Conference on Photovoltaic Energy Conversion (WCPEC) (A Joint Conference of 45th IEEE PVSC, 28th PVSEC & 34th EU PVSEC), pp. 654–659 (2018)
2. Murygin, A.V., Tynchenko, V.S., Laptenok, V.D., Emilova, O.A., Bocharov, A.N.: Complex of automated equipment and technologies for waveguides soldering using induction heating. IOP Conf. Ser. Mater. Sci. Eng. **173**(1), 012023 (2017)
3. Gierth, P., Rebenklau, L., Michaelis, A.: Evaluation of soldering processes for high efficiency solar cells. In: 2012 35th International Spring Seminar on Electronics Technology, pp. 133–137 (2012)
4. Cai, H., Zhao, R.X., Chen, H.M., Wang, S.P.: Study on multiple-frequency IGBT high frequency power supply for induction heating. Proc. CSEE **26**(2), 027 (2006)
5. Wang, Z.D., Luan, X.H., Zhou, Z., Wu, F.S., Zhou, L.Z., Liu, H.: The two-phase flow simulation and experimental research on the formation of solder voids in power module. Microelectron. Reliab. **109**, 113675 (2020)

6. Tian, Y., Wang, L., Anyasodor, G., Xu, Z., Qin, Y.: Heating schemes and process parameters of induction heating of aluminium sheets for hot stamping. Manuf. Rev. **6**, 17 (2019)
7. Nwosu, E., Ewtumo, T., Arogunjo, A., Adeyemi, B.: Analytical modeling with computer simulation validation of an inductive heating system for metals melting application. Iconic Res. Eng. J. **3**(1), 97–103 (2019)
8. Pánek, D., Orosz, T., Kropík, P., Karban, P., Doležel, I.: Reduced-order model based temperature control of induction brazing process. In: 2019 Electric Power Quality and Supply Reliability Conference (PQ) & 2019 Symposium on Electrical Engineering and Mechatronics (SEEM), pp. 1–4 (2019)
9. Eftychiou, M.A., Bergman, T.L., Masada, G.Y.: A detailed thermal model of the infrared reflow soldering process. J. Electron. Packag. **115**(1), 55 (1993)
10. Bukhtoyarov, V.V., Milov, A.V., Tynchenko, V.S., Petrovskiy, E.A., Tynchenko, S.V.: Intelligently informed control over the process variables of oil and gas equipment maintenance. Int. Rev. Autom. Control **12**(2), 59–66 (2019)
11. Panek, D., Karban, P., Dolezel, I.: Calibration of numerical model of magnetic induction brazing. IEEE Trans. Magn. **55**(6), 1–4 (2019)
12. Wei, H.P., Yang, Y.H., Wu, B., Han, B.: Prediction of statistical distribution of vibration-induced solder fatigue failure considering intrinsic variations of mechanical properties of anisotropic Sn-Rich solder alloys. In: 2018 IEEE 68th Electronic Components and Technology Conference (ECTC), pp. 741–747 (2018)
13. Satheesh, A., Kattisseri, M., Vijayan, V.: Numerical estimation of localized transient temperature and strain fields in soldering process. In: 2018 7th Electronic System-Integration Technology Conference (ESTC), pp. 1–5 (2018)
14. Tynchenko, V.S., Murygin, A.V., Petrenko, V.E., Seregin, Y.N., Emilova, O.A.: A control algorithm for waveguide path induction soldering with product positioning. IOP Conf. Ser. Mater. Sci. Eng. **255**(1), 012018 (2017)
15. Petzold, C.: Programming Microsoft Windows with C#. Microsoft Press, Redmond (2002)
16. Hejlsberg, A., Wiltamuth, S., Golde, P.: C# Language Specification. Addison-Wesley Longman Publishing, Boston (2003)
17. Liberty, J.: Programming C#: Building. NET Applications with C. O'Reilly Media, Newton (2005)
18. Schildt, H.: C# 4.0: The Complete Reference. Tata McGraw-Hill Education, New York (2010)
19. Kanaki, K., Kalogiannakis, M.: Introducing fundamental object-oriented programming concepts in preschool education within the context of physical science courses. Educ. Inf. Technol. **23**(6), 2673–2698 (2018). https://doi.org/10.1007/s10639-018-9736-0
20. Strauss, D.: Getting to Know Visual Studio 2019. Apress, Berkeley (2020)
21. Jurgelaitis, M., Čeponienė, L., Čeponis, J., Drungilas, V.: Implementing gamification in a university-level UML modeling course: a case study. Comput. Appl. Eng. Educ. **27**(2), 332–343 (2019)
22. Arora, P.K., Bhatia, R.: Agent-based regression test case generation using class diagram, use cases and activity diagram. Procedia Comput. Sci. **125**, 747–753 (2018)
23. Fauzan, R., Siahaan, D., Rochimah, S., Triandini, E.: Class diagram similarity measurement: a different approach. In: 2018 3rd International Conference on Information Technology, Information System and Electrical Engineering (ICITISEE), pp. 215–219 (2018)

Computing Technologies in Discrete
Mathematics and Decision Making

Evaluation of Efficiency of Using Simple Transformations When Searching for Orthogonal Diagonal Latin Squares of Order 10

Eduard Vatutin[1]([✉]) [iD], Alexey Belyshev[2] [iD], Natalia Nikitina[3] [iD], and Maxim Manzuk[2] [iD]

[1] Southwest State University, 50 let Oktyabrya Street 94, 305040 Kursk, Russia
evatutin@rambler.ru
[2] Internet portal BOINC.ru, Moscow, Russia
[3] Institute of Applied Mathematical Research of Karelian Research Center of RAS, Pushkinskaya Street 11, 185910 Petrozavodsk, Russia

Abstract. The article describes a number of simple transformations (rotation of intercalates, loops, Latin subrectangles, replacement of transversals) in the problem of constructing a collection of orthogonal diagonal Latin squares (ODLS) of order 10 aiming to try to find a triple of pairwise orthogonal diagonal Latin squares of order 10. The averaged time spent on obtaining one canonical form of ODLS using efficient software implementations of square generators based on nested loops and bit arithmetic, on the one hand, and the Euler-Parker method for checking the DLS for the presence of ODLS (together with the composition of the canonizer, if necessary), on the other hand, is 8.3 h using single threaded CPU implementation. It is shown that the application of the indicated transformations in some cases is capable of providing new canonical forms (CFs) of ODLS with significantly lower computational time. The article presents the results of comparing the effectiveness of simple transformations, as well as estimates of the minimum and maximum number of certain structural elements of the DLS (intercalates, loops, Latin subrectangles) depending on its order N (sequences A307163, A307164, A307166, A307167, A307170, A307171, A307841, A307842, A307839, A307840, A287645, A287644 in OEIS). The postprocessor of the found CF of ODLS, working within the framework of the Gerasim@Home volunteer distributed computing project based on this subset of effective simple transformations, allows increasing the output of new CF of ODLS by 10–15% and the required computational costs – by 2–3%.

Keywords: Combinatorics · Diagonal Latin squares · Simple transformations · Euler-Parker method · Integer sequences · OEIS · BOINC

1 Introduction

Latin squares (LS) are a well-known type of combinatorial objects; they are associated with a large number of scientific publications that have both fundamental and applied

© Springer Nature Switzerland AG 2020
V. Jordan et al. (Eds.): HPCST 2020, CCIS 1304, pp. 127–146, 2020.
https://doi.org/10.1007/978-3-030-66895-2_9

significance [1, 2]. The most famous open mathematical problem related to Latin squares is the problem of the existence of a triple of mutually orthogonal Latin squares (MOLS) of order 10. This triple has not yet been found despite many attempts to find, but it has not been theoretically proven that it does not exist. The best approximations to the solution of the indicated problem are a number of pseudo-triples with different orthogonality characteristics, in which some pairs of squares are orthogonal, while others are partially orthogonal in a certain set of cells [3–6].

At different times, various attempts were made to find the desired triple both in an independent form and as part of a more complex combinatorial structure [7] – a graph of LS on a set of binary orthogonality relation. These include: search by reducing the original problem to the problem of Boolean satisfiability (SAT) in order to solve it using specialized solvers [3, 4, 6, 8, 26], construction of orthogonal mates within the framework of the Euler-Parker method [9–11] for various initial squares. These squares provided by different generators: of an arbitrary form, obtained in a pseudo-random way, possessing planar or central symmetry [12, 13], generalized symmetry, generalized symmetry in parastrophic slices, partial generalized symmetry, including in parastrophic slices. Last attempts were connected to the transversal-free search using cell mapping schemes (CMS). In practice, the approach based on the Euler-Parker method shows the greatest efficiency. Within its framework, a set of transversals is constructed and a subset of N disjoint transversals is found among them, and both problems in practice are most effectively implemented by reducing to the problem of exact cover followed by its solution using the dancing links algorithm (DLX) [10, 11]. When working with orthogonal diagonal Latin squares (ODLS), the search efficiency can be further increased within the framework of the procedure called canonization [14].

The team of authors have developed highly efficient software implementations of DLS generators of various types based on nested loops and bit arithmetic operations [15] and a canonizer based on the Euler-Parker method and the dancing link algorithm, which together provide a processing rate of about 8000 DLS/s for a single-threaded software implementation on processor Intel Core i7 4770. According to the results of computations currently performed on the BOINC platform in the Gerasim@Home volunteer distributed computing project [16, 17], to obtain a canonical form (abbr. CF) of the ODLS of order 10 – the lexicographically minimal representative within the corresponding main class of DLS – an average of 30000 s = 8.3 h of CPU computing time is required. The list of findings of the project as of September 2020 contains more than 11.7 million CFs of ODLS of order 10, which, unfortunately, do not include CFs, which forms the target triple of MOLS/MODLS.

It is a well-known fact that for a CF of ODLS it is possible to use a number of transformations which allow one either to obtain new CF of ODLS or new rare combinatorial structures starting from already found ounces (for example, stars of type 1:3 or loop-4) [7]. The description of some of the transformations (for example, the rotation of intercalates), which we will further call simple transformations, is known and can be found on thematic resources, others have been developed by the team of authors and have novelty. For both known and new simple transformations proposed in the article, an analysis of their effectiveness was not carried out before in order to formulate recommendations on the expediency or inexpediency of their practical application in the search for CF of ODLS, which makes this task relevant.

In order to maintain the rigour of the presentation of the article material, we will introduce a number of notations. A Latin square $A = (a_{ij})$, i, $j = \overline{1, N}$ of order N is a square table, the elements (cells) of which are filled with symbols of some alphabet U of cardinality $|U| = N$ (for definiteness, integers 0, 1, ..., $N - 1$) in such a way that in each row and each column the symbols are not repeated. For diagonal Latin squares, an additional restriction is imposed on the uniqueness of the values of elements on the main and secondary diagonals. A Latin rectangle (LR) is a square table with the size of $M \times L$ cells, $1 \leq M \leq N$, $1 \leq L \leq N$, filled with symbols of the alphabet $U' \subseteq U$ of cardinality $|U'| = \max(M, L)$ according to the rules of LS (values in rows and columns are not repeated). A pair of orthogonal LS/DLS (OLS/ODLS) are such squares A and B, that all ordered pairs of values $(a_{ij}, \; b_{ij})$ are different. A transversal T in a Latin square is a set of N cells, one in each column and each row, such that all the values in the cells are different [18–21]. Normalization is such a bijective replacement of the values of the DLS elements when the resulting elements of the first row become ordered in an ascending order.

2   Rotation of Intercalates

The simplest element in the composition of the LSA that allows for change (the so-called rotation) is the intercalate [22, 23], the LS of size 2×2 located at the intersection of some pair of rows i_1 and i_2, $i_1 \neq i_2$, and some pair of columns j_1 and j_2, $j_1 \neq j_2$ of the LS A (Fig. 1).

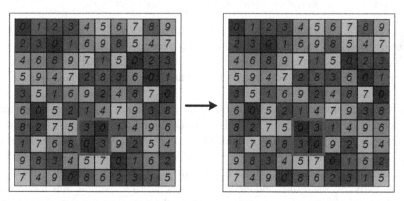

Fig. 1. An example of rotation of the intercalate, providing a new DLS with ODLS from the existing one (in this case, with the same CF).

For LS, this procedure is always performed without violating the definition of the correct LS; for DLS, the rotation of the intercalate can sometimes lead to duplication of values on the diagonals. Sometimes, in the course of performing this transformation using the known ODLS, a new yet unknown CF of ODLS is obtained. If there is a set of K intercalates for the LS undergoing this simple transformation, the rotations can be performed one by one (K possible turns), in pairs (at most $K(K-1)$ possible turns), by 3 intercalates (no more than $K(K-1)(K-2)$ possible turns) etc. During the rotation of one or more intercalates, new (nested) intercalates may appear in the LS (Fig. 2).

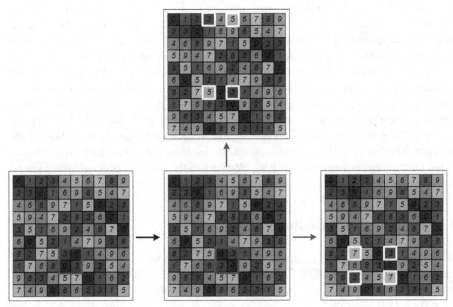

Fig. 2. An example of the emergence of new (nested) intercalates (highlighted in white) as a result of rotation of the intercalate (highlighted in red). (Color figure online)

The LS obtained during the rotation of the intercalates can be analyzed for the presence of ODLS using the Euler-Parker method, or can be subjected to the canonization procedure (the first option requires less computational time, but potentially leads to a smaller number of new CF of ODLS).

The balance between the number of rotations of intercalates and the result of the transformation (the number of new CF of ODLS) was determined empirically in the course of the corresponding computational experiment (see Table 1). It was consisted in obtaining new CF of ODLS according to the known list of CF of ODLS from the Gerasim@Home project, numbering 3.3 million CF of ODLS at the time of the experiment.

Thus, analyzing the results obtained, we can conclude that a simple transformation of the rotation of intercalates is effective for small depths (at most 5 for use in combination with the Euler-Parker method and at most 3 when used in tandem with the canonizer). The search for embedded intercalates significantly increases the computational time spent on the transformation, which is not compensated by the number of new CF of the ODLS, and can only be recommended for rotating 2–3 intercalates in combination with the Euler-Parker method.

Table 1. The results of the computational experiment of rotating D intercalates in various ways (the number of new CF of ODLS; processing rate, seconds per one DLS; time spent on obtaining a new CF of ODLS, seconds). The symbols "+" mark simple transformations that make it possible to obtain new CF of ODLS with less time than when using the scheme "Generator of LS $\rightarrow$ Canonizer".

Depth (D) Transformation	1	2	3	4	5	6	7	8
Rotation of D intercalates, Euler-Parker for correct DLS	+1732 CF	+44190 CF	+6463 CF	+1163 CF	+229012 CF	+57 CF	+48 CF	+27 CF
	0.02 s/DLS	0.06 s/DLS	0.13 s/DLS	0.5 s/DLS	0.3 s/DLS	1 s/DLS	1 s/DLS	1 s/DLS
	38 s/CF (+)	4.5 s/CF (+)	66 s/CF (+)	1418 s/CF (+)	4 s/CF (+)	57900 s/CF	68800 s/CF	122000 s/CF
Rotation of D intercalates, search for new intercalates, Euler-Parker for correct DLS	–	+991 CF	+289 CF	+801 CF	+38 CF	–	–	–
	–	0.1 s/DLS	0.5 s/DLS	40 s/DLS	50 s/DLS	–	–	–
	–	330 s/CF (+)	5709 s/CF (+)	165000 s/CF	4.3 mln. s/CF	–	–	–

The number of structural elements of the DLS that are changed during the execution of each of the considered simple transformations directly affects the cost of computational time. The maximum number of intercalates in the LS is known and is given by the sequence A092237 in the online encyclopedia of integer sequences (OEIS) [22–24]. The values of the sequence are exactly known for the LS of orders $N \leq 9$; for the orders $10 \leq N \leq 12$, only lower constraints are known. For DLS of the order of N, a minimum and maximum number of intercalates at the start of the study were unknown. In order to determine them, a computational experiment was organized, during which all the DLS of a given order were constructed (if it was possible) and the extreme value of the number of intercalates was fixed (Table 3), the found numerical sequences turned out to be new, were tested and added to the OEIS under the numbers A307163 and A307164 (Table 3).

For orders $1 \leq N \leq 8$, the results were obtained by exhaustive search of the entire DLS space. For order $N = 9$, it is difficult to organize a similar computational experiment due to the need for large computational costs. However, it is easy to see that, taking into account the fact that DLS are a subset of LS, for any numerical characteristic X calculated

Table 2. The results of the computational experiment of rotating D intercalates in various ways with the processing of the obtained LS by the canonizer (with the limitation of 1000 LS per 1 initial DLS; otherwise the execution of this simple transformation takes too long).

Depth (D) Transformation	1	2	3	4	5	6	7	8
Rotation of D intercalates, canonizer for the LS	+7705 CF	+38440 CF	+15693 CF	+4354 CF	+5896 CF	+3051 CF	+3868 CF	+4578 CF
	1 s/DLS	9 s/DLS	33 s/DLS	33 s/DLS	40 s/DLS	33 s/DLS	33 s/DLS	33 s/DLS
	520 s/CF (+)	936 c/CF (+)	8411 s/CF (+)	30317 s/CF	27137 s/CF	43264 s/CF	34126 s/CF	28833 s/CF
Rotation of D intercalates, search for new intercalates, canonizer for the LS	–	0	14	12	0	–	–	–
		33 s/DLS	43 s/DLS	43 s/DLS	33 s/DLS			
	–	–	12 mln. s/CF	14 mln. s/CF	–			

Table 3. The minimum and maximum number of intercalates in the DLS of order $1 \leq N \leq 10$.

N	Min (A307163)	Max (A307164)
1	0	0
2	0	0
3	0	0
4	12	12
	0 1 2 3	0 1 2 3
	3 2 1 0	3 2 1 0
	1 0 3 2	1 0 3 2
	2 3 0 1	2 3 0 1
5	0	4
	0 1 2 3 4	0 1 2 3 4
	4 2 3 0 1	4 2 0 1 3
	3 4 1 2 0	1 4 3 2 0
	1 3 0 4 2	3 0 1 4 2
	2 0 4 1 3	2 3 4 0 1
6	9	9
	0 1 2 3 4 5	0 1 2 3 4 5
	4 2 5 0 3 1	4 2 5 0 3 1
	3 5 1 2 0 4	3 5 1 2 0 4
	5 3 0 4 1 2	5 3 0 4 1 2
	2 4 3 1 5 0	2 4 3 1 5 0
	1 0 4 5 2 3	1 0 4 5 2 3
7	0	30
	0 1 2 3 4 5 6	0 1 2 3 4 5 6
	4 2 6 0 5 1 3	4 2 6 5 0 1 3
	3 5 1 6 0 4 2	3 6 1 0 5 2 4
	5 6 3 4 1 2 0	6 3 5 4 1 0 2
	6 4 5 2 3 0 1	1 5 3 2 6 4 0
	1 3 0 5 2 6 4	5 0 4 6 2 3 1
	2 0 4 1 6 3 5	2 4 0 1 3 6 5
8	0	112
	0 1 2 3 4 5 6 7	0 1 2 3 4 5 6 7
	3 2 5 1 6 7 0 4	4 2 1 7 0 6 5 3
	6 4 1 0 7 2 5 3	5 7 3 2 6 0 4 1
	2 7 3 4 5 0 1 6	7 5 6 4 3 1 2 0
	7 5 0 6 3 4 2 1	6 3 7 1 5 4 0 2
	5 0 4 7 1 6 3 2	1 0 4 6 2 7 3 5
	4 3 6 5 2 1 7 0	3 6 5 0 7 2 1 4
	1 6 7 2 0 3 4 5	2 4 0 5 1 3 7 6

(*continued*)

Table 3. (*continued*)

	0	72
	0 1 2 3 4 5 6 7 8	0 1 2 3 4 5 6 7 8
	4 8 3 6 7 2 1 0 5	6 8 3 4 7 2 0 5 1
	7 6 1 4 5 8 2 3 0	5 7 6 1 8 0 2 4 3
9	5 4 7 2 6 1 0 8 3	7 3 8 2 5 4 1 0 6
	1 5 0 7 3 6 8 4 2	2 4 0 8 1 3 5 6 7
	3 0 6 5 8 4 7 2 1	4 5 1 6 0 7 3 8 2
	2 3 4 8 0 7 5 1 6	8 6 7 5 3 1 4 2 0
	8 7 5 1 2 0 3 6 4	1 0 5 7 2 6 8 3 4
	6 2 8 0 1 3 4 5 7	3 2 4 0 6 8 7 1 5
	≤ 2	≥ 75
	0 1 2 3 4 5 6 7 8 9	0 1 2 3 4 5 6 7 8 9
	1 2 0 4 5 7 3 9 6 8	1 2 0 4 7 9 5 3 6 8
	4 6 5 7 9 8 0 2 1 3	4 9 5 1 6 7 0 8 2 3
	3 7 1 8 2 0 4 6 9 5	9 5 7 6 8 1 3 2 4 0
10	9 3 6 0 7 1 5 8 4 2	7 0 8 2 3 4 9 5 1 6
	5 9 7 6 8 4 1 3 2 0	6 3 4 9 0 8 2 1 5 7
	2 8 4 5 6 3 9 0 7 1	8 4 9 5 1 0 7 6 3 2
	8 4 3 9 0 6 2 1 5 7	3 6 1 0 9 2 8 4 7 5
	6 0 8 2 1 9 7 5 3 4	5 7 3 8 2 6 1 0 9 4
	7 5 9 1 3 2 8 4 0 6	2 8 6 7 5 3 4 9 0 1

for them, the following inequality is valid:

$$0 \leq X_{\min}^{LS}(N) \leq X_{\min}^{DLS}(N) \leq X_{\max}^{DLS}(N) \leq X_{\max}^{LS}(N).$$

Taking into account the known numerical series in the OEIS for the number of intercalates, the inequality takes the following form:

$$0 \leq X_{\min}^{LS}(N) \leq A307163(N) \leq A307164(N) \leq A092237(N).$$

In the course of a limited enumeration for the dimension $N = 9$, we managed to find two DLS, for which the number of intercalates is 0 and 72, respectively. The first value coincides with the lower limit, and the second value with the upper one (for LS). For dimensionality $N = 10$, at present, only the indication of the upper and lower boundaries is possible. They were obtained by analysing the squares included in the list of CF of ODLS of order 10, found in the Gerasim@Home project.

3   Concept and Rotation of Loops

The concept of intercalates can be generalized to loops within a LS. We take any row of the square and choose any two values in it. Let these be, for the square shown in Fig. 3 at the top in the middle, the values 4 and 1 in cells [1] and [1, 3].

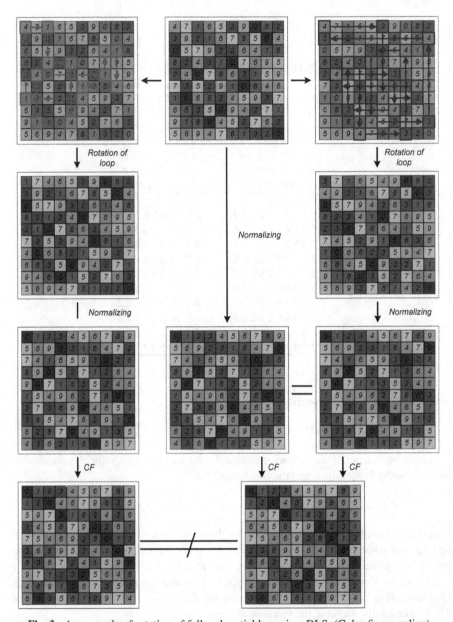

Fig. 3. An example of rotation of full and partial loops in a DLS. (Color figure online)

Further, alternately changing the direction of movement vertically and horizontally, it is necessary to find the same values:

$$L[1, 1] = 4 \rightarrow L[1, 3] = 1 \rightarrow L[4, 3] = 4 \rightarrow L[4, 5] = 1 \rightarrow \ldots \rightarrow L[1, 7] = 1 \rightarrow L[1, 1] = 4.$$

In the course of such a traversal, a closed loop in the LS was obtained. Replacing the values 1 by 4 (the so-called rotation of the loop) in its composition, one can get a new correct LS, which is quite close (by Hamming) to the original LS and, with a high probability, will have similar properties (for example, the presence of an ODLS).

The maximum loop length (in LS cells) is $2N$, such loop will be called full. A loop of shorter length will be called partial. The minimum loop length is 4, and the loop itself is an intercalate. Thus, the loops in the LS can be considered as the possible generalizations of intercalates.

The rotation of the full loop is not interesting, since it is very easy to make sure that the obtained LS after normalization will be exactly equal to the original one (this situation is shown in Fig. 3 on the right; one of the full loops is highlighted in blue). For the same reason, the rotation of all the loops formed by the pair of values v_1 and v_2 is not interesting. Rotation of a loop that is not full gives another DLS from a different main class with other properties that are likely to be close to the properties of the original DLS (this situation is illustrated in Fig. 3 on the left for the loop highlighted in red).

In order to develop a new simple transformation, similar to that considered above with intercalates, it is possible to organize a rotation of 1, 2, 3 and a larger number of loops. In order to assess the effectiveness of this transformation, a corresponding computational experiment was organized with the same initial conditions as the previous one; its results are shown in Table 4.

Table 4. Results of the computational experiment of rotating D partial loops.

Depth (D) Transformation	1	2	3	4	5
Rotation of D loops, Euler-Parker for correct DLS	+42 CF	+28 CF	+51 CF	+38 CF	+8 CF
	0.07 s/DLS	0.4 s/DLS	2 s/DLS	8 s/DLS	40 s/DLS
	5500 c/CF	47 thous.s/CF	129 thous.s/CF	694 thous.s/CF	16 mln. s/CF
Rotation of D loops, canonizer for LS	+2400 CF	+72 CF	+0 CF	+0 CF	+0 CF
	5 s/DLS	56 s/DLS	58 s/DLS	58 s/DLS	40 s/DLS
	8333 s/CF	3.1 mln. s/CF	–	–	–

The obtained results allow us to conclude that the transformation of the rotation of the loops does not lead to obtaining a large number of new CF of ODLS of order 10 and can be limitedly recommended for practical use only for the case of rotation of one loop, where it demonstrates relative efficiency.

The dependencies of the minimum and maximum number of loops on the order N of DLS, were not presented before this study. For the orders $1 \leq N \leq 7$, they were determined by a complete enumeration of all DLS. For dimensionality $N = 8$ this required the development of a specialized CF DLS generator based on X-based diagonal fillings and ESODLS schemes. (The numerical characteristics indicated in the article are invariants of the main class of DLS, therefore, to search for DLS with an extreme value of the selected characteristic, it is enough to analyze one DLS from each main class, which reduces the computational costs for organizing the search by 2-3 orders of magnitude).

Table 5. The minimum and maximum number of loops in a DLS of order N.

N	Min (A307166)	Max (A307167)
1	1	1
	0	0
2	0	0
3	0	0
4	12	12
	0 1 2 3	0 1 2 3
	3 2 1 0	3 2 1 0
	1 0 3 2	1 0 3 2
	2 3 0 1	2 3 0 1
5	10	14
	0 1 2 3 4	0 1 2 3 4
	4 2 3 0 1	4 2 0 1 3
	3 4 1 2 0	1 4 3 2 0
	1 3 0 4 2	3 0 1 4 2
	2 0 4 1 3	2 3 4 0 1
6	27	27
	0 1 2 3 4 5	0 1 2 3 4 5
	4 2 5 0 3 1	4 2 5 0 3 1
	3 5 1 2 0 4	3 5 1 2 0 4
	5 3 0 4 1 2	5 3 0 4 1 2
	2 4 3 1 5 0	2 4 3 1 5 0
	1 0 4 5 2 3	1 0 4 5 2 3
7	21	53
	0 1 2 3 4 5 6	0 1 2 3 4 5 6
	4 2 6 0 5 1 3	4 2 6 5 0 1 3
	3 5 1 6 0 4 2	3 6 1 0 5 2 4
	5 6 3 4 1 2 0	6 3 5 4 1 0 2
	6 4 5 2 3 0 1	1 5 3 2 6 4 0
	1 3 0 5 2 6 4	5 0 4 6 2 3 1
	2 0 4 1 6 3 5	2 4 0 1 3 6 5
8	40	112
	0 1 2 3 4 5 6 7	0 1 2 3 4 5 6 7
	1 7 6 4 5 3 0 2	6 7 4 5 2 3 0 1
	6 3 1 5 0 2 7 4	3 2 1 0 7 6 5 4
	3 0 4 2 1 7 5 6	5 4 7 6 1 0 3 2
	7 2 5 6 3 4 1 0	1 0 3 2 5 4 7 6
	5 4 3 0 7 6 2 1	7 6 5 4 3 2 1 0
	2 5 7 1 6 0 4 3	2 3 0 1 6 7 4 5
	4 6 0 7 2 1 3 5	4 5 6 7 0 1 2 3

Table 6. The minimum and maximum number of partial loops in DLS of order N.

N	Min (A307170)	Max (A307171)
1	0	0
2	0	0
3	0	0
4	12	12
	0 1 2 3	0 1 2 3
	3 2 1 0	3 2 1 0
	1 0 3 2	1 0 3 2
	2 3 0 1	2 3 0 1
5	0	8
	0 1 2 3 4	0 1 2 3 4
	4 2 3 0 1	4 2 0 1 3
	3 4 1 2 0	1 4 3 2 0
	1 3 0 4 2	3 0 1 4 2
	2 0 4 1 3	2 3 4 0 1
6	21	21
	0 1 2 3 4 5	0 1 2 3 4 5
	4 2 5 0 3 1	4 2 5 0 3 1
	3 5 1 2 0 4	3 5 1 2 0 4
	5 3 0 4 1 2	5 3 0 4 1 2
	2 4 3 1 5 0	2 4 3 1 5 0
	1 0 4 5 2 3	1 0 4 5 2 3
7	0	53
	0 1 2 3 4 5 6	0 1 2 3 4 5 6
	4 2 6 0 5 1 3	4 2 6 5 0 1 3
	3 5 1 6 0 4 2	3 6 1 0 5 2 4
	5 6 3 4 1 2 0	6 3 5 4 1 0 2
	6 4 5 2 3 0 1	1 5 3 2 6 4 0
	1 3 0 5 2 6 4	5 0 4 6 2 3 1
	2 0 4 1 6 3 5	2 4 0 1 3 6 5
8	24	112
	0 1 2 3 4 5 6 7	0 1 2 3 4 5 6 7
	1 7 6 4 5 3 0 2	6 7 4 5 2 3 0 1
	6 3 1 5 0 2 7 4	3 2 1 0 7 6 5 4
	3 0 4 2 1 7 5 6	5 4 7 6 1 0 3 2
	7 2 5 6 3 4 1 0	1 0 3 2 5 4 7 6
	5 4 3 0 7 6 2 1	7 6 5 4 3 2 1 0
	2 5 7 1 6 0 4 3	2 3 0 1 6 7 4 5
	4 6 0 7 2 1 3 5	4 5 6 7 0 1 2 3

The resulting numerical sequences are new, they have been tested and have been added to the OEIS under numbers A307166, A307167, A307170 and A307171.

Each partial loop is a loop in a diagonal Latin square by definition. Similarly, by definition, each intercalate is a partial loop of length 4. These simple statements allow us to establish a number of relationships between the values of the numerical series associated with loops and intercalates and calculated earlier:

$$0 \leq A307166(N) \leq A307167(N),$$

$$0 \leq A307170(N) \leq A307171(N),$$

$$0 \leq A307163(N) \leq A307170(N) \leq A307166(N),$$

$$A307164(N) \leq A307171(N) \leq A307167(N).$$

Similar values for LS are currently unknown (Tables 5 and 6).

4 The Concept and "Rotation" of Latin Subrectangles in DLS

Intercalates are of size 2×2 by definition LS. It is easy to make sure that in the composition of the squares it is possible to find a LS of a different size (3×3, 4×4 etc.), and instead of the found LS, one can put any other LS of the same size from the same values ("rotate" the found LS). In order to avoid terminological confusion, such "small" LS included in the "large" ones will be called sub-squares.

For example, instead of the sub-square located in the upper left corner (see Fig. 4a)

$$\begin{matrix} 0 & 1 & 2 \\ 1 & 2 & 0 \\ 2 & 0 & 1 \end{matrix}$$

one can put any "small" LS, for example,

$$\begin{matrix} 0 & 2 & 1 \\ 2 & 1 & 0 \\ 1 & 0 & 2 \end{matrix}$$

or

$$\begin{matrix} 1 & 0 & 2 \\ 0 & 2 & 1 \\ 2 & 1 & 0 \end{matrix}$$

and so on, while the "large" LS remains correct. For DLS, this transformation sometimes leads to the appearance of duplicate values on the diagonals, but this is easily tracked and does not happen often.

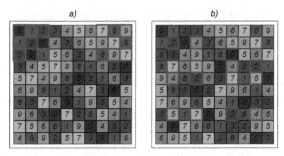

Fig. 4. Examples of DLS, including sub-squares of size 3×3 and 4×4.

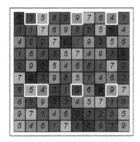

Fig. 5. An example of a DLS, including a Latin subrectangle of size 2×4.

Similarly, in the DLS one can search for Latin subrectangles – such sets of M rows $\{i_1, i_2, \ldots, i_M\}$ and L columns $\{j_1, j_2, \ldots, j_L\}$ at the intersection of which there is a Latin rectangle of size $M \times L$ formed from $\max(M, L)$ values (see Fig. 5, elements of one of the Latin rectangles are highlighted in white).

Within the framework of the found Latin rectangle, one can rearrange the rows or columns of a smaller dimension; in this case, as with the transformations of the "small" LS discussed above, the "large" LS remains a correct LS, and sometimes, but not often, duplicate values may appear in the DLS on the diagonals. Latin rectangles, by definition, are all pairs of cells of a square of size $Y \times 1$ and $Y \times 1, 1 \leq Y \leq N$. Also, Latin subrectangles can have size $Y \times N$ and $N \times Y$, $Y < N$, i.e. in fact, any combination of rows and columns of the DLS falls under the definition of a correct Latin rectangle, as well as the "large" square itself – under the definition of a "small" subLS. These Latin rectangles and squares will be called trivial – they are not interesting in the context of the simple transformations considered in the article. Nontrivial LS and LR can be "rotated", for which it is necessary to build a list of all nontrivial LS and LR for a given DLS, and then, as with the intercalates and loops discussed above, "rotate" a combination of 1, 2, 3 etc. more LS/LR. In order to assess the effectiveness of this transformation, a corresponding computational experiment was organized with the same initial conditions as the experiments considered above, its results are shown in Table 7.

Table 7. Results of the computational experiment of "rotating" *DLS/LR*.

Depth (D) Transformation	1	2	3	4	5
Rotation of D Latin rectangles/squares, Euler-Parker for correct DLS	+16252 CF	+703 CF	+68 CF	+7 CF	+10 CF
	0.5 s/DLS	3 s/DLS	3 s/DLS	3 s/DLS	3 s/DLS
	101 s/CF (+)	14 thous. s/CF	145 thous. s/CF	1.4 mln. s/CF	1 mln. s/CF
Rotation D Latin rectangles/squares, canonizer for LS	+170826 CF	+10464 CF	+17366 CF	+224398 CF	–
	14 s/DLS	18 s/DLS	34 s/DLS	33 s/DLS	–
	327 s/CF (+)	6880 s/CF	7831 s/CF	588 s/CF (+)	–

The obtained results allow us to conclude that the most effective transformations are the "rotation" of one LS/LR with the processing of the resulting square by the Euler-Parker method and the "rotation" of one and four LS/LR when processing by the canonizer.

The dependences of the minimum and maximum number of Latin subrectangles on the order of DLS N were unknown before this study. They were established in the course of corresponding computational experiments, the results of which are presented in Tables 8 and 9. For dimensions $1 \leq N \leq 7$, the indicated numerical characteristics were determined by an exhaustive search of all possible squares of a given dimension, for dimension $N = 8$ in order to reduce time costs to reasonable values, a CF DLS generator based on X-based diagonal fillings and ESODLS schemes was used. The found numerical series were tested and were added to the OEIS under the numbers A307839, A307840, A307841 and A307842.

By analogy with inequalities for loops, we can formulate a number of inequalities for Latin subrectangles in DLS. Every nontrivial subrectangle is a subrectangle in the DLS by definition. Similarly, by definition, each intercalate is a nontrivial 2×2 subrectangle (except for the dimension $N = 2$, where it will be trivial, but there is no DLS of this dimension). These simple statements allow us to establish a number of relationships between the values of numerical series associated with subrectangles and intercalates:

$$0 \leq A307839(N) \leq A307840(N),$$

$$0 \leq A307841(N) \leq A307842(N)$$

$$0 \leq A307163(N) \leq A307841(N) \leq A307839(N),$$

$$A307164(N) \leq A307842(N) \leq A307840(N).$$

Table 8. The minimum and maximum number of nontrivial Latin subrectangles in the DLS of order N.

N	Min (A307841)	Max (A307842)
1	0	0
2	0	0
3	0	0
4	12	12
	0 1 2 3	0 1 2 3
	3 2 1 0	3 2 1 0
	1 0 3 2	1 0 3 2
	2 3 0 1	2 3 0 1
5	0	12
	0 1 2 3 4	0 1 2 3 4
	4 2 3 0 1	4 2 3 0 1
	3 4 1 2 0	3 4 1 2 0
	1 3 0 4 2	1 3 0 4 2
	2 0 4 1 3	2 0 4 1 3
6	51	51
	0 1 2 3 4 5	0 1 2 3 4 5
	4 2 5 0 3 1	4 2 5 0 3 1
	3 5 1 2 0 4	3 5 1 2 0 4
	5 3 0 4 1 2	5 3 0 4 1 2
	2 4 3 1 5 0	2 4 3 1 5 0
	1 0 4 5 2 3	1 0 4 5 2 3
7	0	151
	0 1 2 3 4 5 6	0 1 2 3 4 5 6
	4 2 6 0 5 1 3	4 2 6 5 0 1 3
	3 5 1 6 0 4 2	3 6 1 0 5 2 4
	5 6 3 4 1 2 0	6 3 5 4 1 0 2
	6 4 5 2 3 0 1	1 5 3 2 6 4 0
	1 3 0 5 2 6 4	5 0 4 6 2 3 1
	2 0 4 1 6 3 5	2 4 0 1 3 6 5
8	36	924
	0 1 2 3 4 5 6 7	0 1 2 3 4 5 6 7
	5 7 4 6 3 2 0 1	6 7 4 5 2 3 0 1
	4 3 5 7 0 1 2 6	3 2 1 0 7 6 5 4
	3 5 0 1 6 7 4 2	5 4 7 6 1 0 3 2
	7 0 6 5 2 4 1 3	1 0 3 2 5 4 7 6
	1 2 3 4 5 6 7 0	7 6 5 4 3 2 1 0
	6 4 1 2 7 0 3 5	2 3 0 1 6 7 4 5
	2 6 7 0 1 3 5 4	4 5 6 7 0 1 2 3

Table 9. The minimum and maximum number of Latin subrectangles in DLS of order N.

N	Min (A307839)	Max (A307840)
1	1	1
2	0	0
3	0	0
4	137	137
	0 1 2 3	0 1 2 3
	3 2 1 0	3 2 1 0
	1 0 3 2	1 0 3 2
	2 3 0 1	2 3 0 1
5	336	348
	0 1 2 3 4	0 1 2 3 4
	4 2 3 0 1	4 2 0 1 3
	3 4 1 2 0	1 4 3 2 0
	1 3 0 4 2	3 0 1 4 2
	2 0 4 1 3	2 3 4 0 1
6	884	884
	0 1 2 3 4 5	0 1 2 3 4 5
	4 2 5 0 3 1	4 2 5 0 3 1
	3 5 1 2 0 4	3 5 1 2 0 4
	5 3 0 4 1 2	5 3 0 4 1 2
	2 4 3 1 5 0	2 4 3 1 5 0
	1 0 4 5 2 3	1 0 4 5 2 3
7	1968	2119
	0 1 2 3 4 5 6	0 1 2 3 4 5 6
	4 2 6 0 5 1 3	4 2 6 5 0 1 3
	3 5 1 6 0 4 2	3 6 1 0 5 2 4
	5 6 3 4 1 2 0	6 3 5 4 1 0 2
	6 4 5 2 3 0 1	1 5 3 2 6 4 0
	1 3 0 5 2 6 4	5 0 4 6 2 3 1
	2 0 4 1 6 3 5	2 4 0 1 3 6 5
8	4545	5433
	0 1 2 3 4 5 6 7	0 1 2 3 4 5 6 7
	5 7 4 6 3 2 0 1	6 7 4 5 2 3 0 1
	4 3 5 7 0 1 2 6	3 2 1 0 7 6 5 4
	3 5 0 1 6 7 4 2	5 4 7 6 1 0 3 2
	7 0 6 5 2 4 1 3	1 0 3 2 5 4 7 6
	1 2 3 4 5 6 7 0	7 6 5 4 3 2 1 0
	6 4 1 2 7 0 3 5	2 3 0 1 6 7 4 5
	2 6 7 0 1 3 5 4	4 5 6 7 0 1 2 3

5 Replacement of Transversals

Another structural element of DLS, suitable for replacement within the framework of the next simple transformation, is transversals [18–20]. The essence of a simple transformation based on them boils down to resetting the values of the elements included in the selected subset of transversals, followed by the completion of a partially filled square and processing the obtained LS/DLS by the Euler-Parker method or the canonizer (Fig. 6).

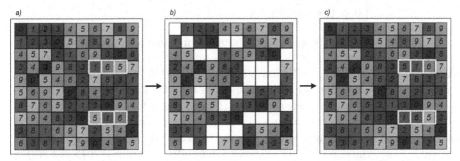

Fig. 6. Illustration of a simple transformation of replacing transversals: the original square (a), a partially filled square with three reset transversals (b) and the result of one of the completions leading to a new square (c). In the given example, this transformation is equivalent to "rotating" the Latin subrectangle of size 2×3 (highlighted in yellow). (Color figure online)

In order to analyze the efficiency of this transformation, a computational experiment was organized with conditions similar to the experiments considered above. When performing this simple transformation, most of the resulting squares are LS, not DLS; when processing them by the Euler-Parker method, the main part of the computational time is spent on checking the correctness of the square, and not on finding the ODLS to it. Therefore, this type of transformation can be considered ineffective; it was not possible to obtain new CF of ODLS with its use. The results of the experiment with the processing of the obtained LS by the canonizer are shown in Table 10.

Table 10. Results of the computational experiment of replacing D transversals.

Depth (D) Transformation	1	2	3	4	5
Replacement of D transversals	–	0	+9949 CF	+2260 CF	+2078 CF
		–	16082 s/CF	4424 s/CF (+)	4812 s/CF (+)

The results of the computational experiment show that the transformations for the replacement of 4 and 5 transversals can be considered relatively efficient, providing a relatively small number of new CF of ODLS.

The number of transversals in the DLS is known; it was calculated in [25] for dimensions $1 \leq N \leq 8$. For the dimension $N = 10$ for the square from [14], the lower constraint is known (≥ 5504); for other dimensions neither the exact value nor the lower constraints were known at the time of the study. For transversals, as well as for intercalates, the following relations can be formulated:

$$0 \leq A287645(N) \leq A287644(N) \leq A090741(N),$$

$$\begin{cases} A091323\left(\frac{N-1}{2}\right) \leq A287645(N), & N \bmod 2 = 1, \\ 0 \leq A287645(N), & N \bmod 2 = 0. \end{cases}$$

The first of them is an upper constraint on the maximum number of transversals in the DLS (sequence A287644) from the side of a similar characteristic in the LS (A090741), the second is a similar lower constraint (A091323 for odd orders and 0 for even) on the minimum number of transversals in the DLS (A287645). In the course of the computational experiments performed (limited search for DLSs of order 9), two DLSs were found with the minimum and maximum possible number of transversals in the DLS, which coincides with similar known characteristics for the LS, which makes it possible to determine previously unknown values $a(9) = 68$ in A287645 and $a(9) = 2241$ in A287644 (Table 11).

Table 11. The minimum and maximum number of transversals in DLS of order 9.

Min (A287645)	Max (A287644)
68	2241
0 1 2 3 4 5 6 7 8	0 1 2 3 4 5 6 7 8
4 8 5 2 1 7 3 0 6	7 8 4 0 6 1 2 3 5
2 5 1 7 0 6 4 8 3	4 3 1 2 8 7 5 6 0
3 2 6 4 8 1 7 5 0	1 6 7 5 3 4 0 8 2
5 3 7 6 2 0 8 1 4	5 4 0 8 7 2 3 1 6
1 6 8 5 7 3 0 4 2	8 2 3 1 0 6 7 5 4
7 4 3 0 6 8 5 2 1	3 5 6 7 2 8 4 0 1
8 7 0 1 3 4 2 6 5	6 0 8 4 5 3 1 2 7
6 0 4 8 5 2 1 3 7	2 7 5 6 1 0 8 4 3

6   Conclusion

As a result of the study, a number of new simple transformations was developed, and the efficiency of the new and previously known ones was analyzed in the search for CF of ODLS of order 10. Basing on the computational experiments, one can conclude that the rotation of intercalates is effective at a conversion depth of at most 5 in combination with the Euler-Parker method and at most 3 in combination with the canonizer. The rotation of the loops does not show high efficiency and can be limitedly recommended for practical use for the rotation of a single loop. "Rotation" of Latin subrectangles and sub-squares in the DLS is effective for 1 LS/LR in combination with the Euler-Parker method and for 1 and 4 LS/LR in combination with the canonizer. Replacement of transversals is relatively efficient for 4 and 5 transversals in combination with the canonizer. In addition to obtaining just new CF of ODLS, the use of the simple transformations considered in the article made it possible to obtain a number of rare combinatorial structures: 68 CF of loop-4 and 4 CF of line-4 by rotating intercalates, 4 CF of loop-4 by rotating loops, 4 CF of star 1:3 and 4 CF of line-4 by "rotation" of Latin subrectangles. It should be assumed that the efficiency of applying simple transformations both for the OLS and for other orders N of the ODLS will differ. In order to prove the effectiveness of using simple transformations in other conditions, a series of computational experiments will be required, similar to that considered in this article.

At present, a subset of practically effective simple transformations is integrated into the computational module working within the framework of the Gerasim@Home volunteer distributed computing project, which is engaged in building a collection of CF of ODLS of order 10. Processing of DLS and LS of various types is carried out according to the scheme "Generator of LS/DLS → Canonizer → Postprocessor". Simple transformations are performed as part of a postprocessor on the client side for the found CF of ODLS, which insignificantly (about 2–3%) increases the computational time spent on the execution of the computational task and leads to a final 10–15% increase in the number of CF of ODLS. At the present time, CF of ODLS of order 10 form 24 types of different combinatorial structures, which unfortunately do not include the triple of MOLS/MODLS.

Acknowledgments. The article partially supported by RFBR, grant number 18-07-00628. The authors of the article would like to thank all the crunchers who took part in the Gerasim@Home volunteer distributed computing project, as well as citerra and SerVal members (the Russia Team) of the BOINC.ru Internet portal for their help in organizing computational experiments and constructive discussion of their details.

References

1. Colbourn, C.J., Dinitz, J.H.: Handbook of Combinatorial Designs, 2nd edn. Chapman & Hall/CRC, London (2006)
2. Keedwell, A.D., Dénes, J.: Latin Squares and Their Applications. Elsevier, Heidelberg (2015). https://doi.org/10.1016/c2014-0-03412-0
3. Zaikin, O.S., Kochemazov, S.E., Belorechev, I.D.: The search for pairs of orthogonal diagonal Latin squares of order 10 in the volunteer distributed computing project SAT@Home. In: Parallel Computational Technologies (PCT 2015), pp. 157–165 (2015). (in Russian)
4. Zaikin, O., Kochemazov, S.: The search for systems of diagonal Latin squares using the SAT@Home project. Int. J. Open Inf. Technol. 3(11), 4–9 (2015)
5. Brouwer, A.E.: Four MOLS of order 10 with a hole of order 2. J. Statist. Plann. Inference **10**, 203–205 (1984)
6. Zaikin, O., Zhuravlev, A., Kochemazov, S., Vatutin, E.: On the construction of triples of diagonal latin squares of order 10. Electronic Notes in Discrete Mathematics **54C**, 307–312 (2016). https://doi.org/10.1016/j.endm.2016.09.053

7. Vatutin, E.I., Titov, V.S., Zaikin, O.S., Kochemazov, S.E., Manzuk, M.O., Nikitina, N.N.: Orthogonality-based classification of diagonal Latin squares of order 10. In: CEUR Workshop Proceedings, vol. 2267, pp. 282–287 (2018)
8. Zaikin, O.S., Kochemazov, S.E.: The search for pairs of orthogonal diagonal Latin squares of order 10 in the volunteer distributed computing project SAT@Home. Bull. South Ural State Univ. Ser. Comput. Math. Softw. Eng. 4(3), 95–108 (2015). (in Russian)
9. Parker, E.T.: Orthogonal Latin squares. Proc. Natl. Acad. Sci. U.S.A. 45(6), 859–862 (1959)
10. Knuth, D.E.: Dancing links. In: Millenial Perspectives in Computer Science, pp. 187–214 (2000)
11. Knuth, D.E.: The Art of Computer Programming. Volume 4A – Combinatorial Algorithms, Part 1. Addison-Wesley, Reading (2011). xv+883 pp.
12. Vatutin, E.I., Kochemazov, S.E., Zaikin, O.S.: On some features of symmetric diagonal Latin squares. In: CEUR Workshop Proceedings, vol. 1940, pp. 74–79 (2017)
13. Vatutin, E.I., Kochemazov, S.E., Zaikin, O.S., Manzuk, M.O., Nikitina, N.N., Titov, V.S.: Central symmetry properties for diagonal Latin squares. Probl. Inf. Technol. 2, 3–8 (2019). https://doi.org/10.25045/jpit.v10.i2.01
14. Brown, J.W., Cherry, F., Most, L., Most, M., Parker, E.T., Wallis, W.D.: Completion of the spectrum of orthogonal diagonal Latin squares. Lect. Notes Pure Appl. Math. 139, 43–49 (1992)
15. Kochemazov, S., Zaikin, O., Vatutin, E., Belyshev, A.: Enumerating Diagonal latin squares of order up to 9. J. Integer Seq. 23(1), 20.1.2 (2020)
16. Anderson, David P.: BOINC: a platform for volunteer computing. J. Grid Comput. 18(1), 99–122 (2019). https://doi.org/10.1007/s10723-019-09497-9
17. Gerasim@Home volunteer distributed computing project. http://gerasim.boinc.ru. Accessed 20 Oct 2020
18. Bedford, D.: Transversals in the Cayley tables of the non-cyclic groups of order 8. Eur. J. Comb. 12, 455–458 (1991)
19. McKay, B.D., McLeod, J.C., Wanless, I.M.: The number of transversals in a Latin square. Des. Codes Cryptogr. 40, 269–284 (2006)
20. Cavenagh, N.J., Wanless, I.M.: On the number of transversals in Cayley tables of cyclic groups. Disc. Appl. Math. 158, 136–146 (2010)
21. Potapov, V.N.: On the number of transversals in Latin squares. arXiv:1506.01577 [math.CO] (2015)
22. Bean, R.: Critical sets in Latin squares and associated structures. Ph.D. thesis, The University of Queensland (2001)
23. Heinrich, K., Wallis, W.: The maximum number of intercalates in a Latin square. In: Combinatorial Mathematics VIII, Proceedings of 8th Australian Conference Combinatorics, pp. 221–233 (1980)
24. Sloane, N.J.A.: The on-line encyclopedia of integer sequences. https://oeis.org. Accessed 20 Oct 2020
25. Vatutin, E.I., Kochemazov, S.E., Zaikin, O.S., Valyaev, S.Yu.: Enumerating the transversals for diagonal Latin squares of small order. In: CEUR Workshop Proceedings, vol. 1973, pp. 6–14 (2017)
26. Zaikin, O.S., Kochemazov, S.E., Semenov, A.A.: SAT-based search for systems of diagonal Latin squares in volunteer computing project SAT@home. In: 39th International Convention on Information and Communication Technology, Electronics and Microelectronics (MIPRO 2016), pp. 277–281 (2016)

Implementation of the Extrapolation Method of Expert Assessments in Selection Problems

Svetlana Chernyaeva$^{(\boxtimes)}$ ⓘ, Lyudmila Korobova, Maxim Ivliev ⓘ, Irina Tolstova ⓘ, Boris Nikitin ⓘ, and Irina Matytsina ⓘ

Voronezh State University of Engineering Technologies (VSUET), Revolution Avenue 19, 394036 Voronezh, Russia

`chernsv1978@gmail.com, lyudmila_korobova@mail.ru, max1m@mail.ru, irin2102ka@mail.ru, nbe6419@gmail.com, irina210390@mail.ru`

Abstract. The article describes an algorithm for calculating objective estimates of alternatives and their statistical service characteristics based on elements of the matrix of paired comparisons. The method of applying the method of extrapolation of expert assessments to build a rating of various institutions, including educational organizations, is proposed. The rating task is considered as a multi-criteria decision-making task. A method for finding estimates of the coefficients of a function of a generalized criterion of a given structure is described. When analyzing the performance of various organizations, it is important to determine the focus of the study — whether it will be conducted on the entire population of research objects or only on a separate subgroup. Most indicators cannot be used without taking into account the specifics of the object under study. Research shows that some properties and growth dynamics differ, but not all properties are represented for all objects. You should also be careful about the observed trends, since fluctuations can be associated not only with changes in the quality of functioning of the objects under consideration, but also with data collection. A possible strategy is to find limiting distributions, which allows us to draw cautious conclusions about the stability (or lack thereof) of the observed process. A software product has been developed that allows: to perform automated calculation of performance parameters of normalized values of indicators for each evaluation criterion; to analyze the training sample; to rank the objects under study based on the results of the training sample.

Keywords: Hierarchy analysis method · Expert evaluation extrapolation method · Utility function · Generic criterion · Statistical estimates · Performance analysis

1 Introduction

There is a wide variety of methods that help the researcher to select the best option among the many possible outcomes or available alternatives. Tasks where these methods are used relate to selection and decision tasks. These include the multi-criteria alternatives variety building a rating task. Such tasks are evaluated in different categories. Number

© Springer Nature Switzerland AG 2020
V. Jordan et al. (Eds.): HPCST 2020, CCIS 1304, pp. 147–161, 2020.
https://doi.org/10.1007/978-3-030-66895-2_10

of outcomes or alternatives and criteria number should be highlighted among them. Both categories have a significant impact on the algorithmic solution scheme and the computational selection problem complexity. With a large alternatives number (hundreds or thousands), choosing the only right alternative is not possible; therefore, there is a need for clear algorithms. The number of criteria used in the selection tasks will greatly complicate the solution problem. The difficulty arises in the fact that it is impossible to develop a solution based on only one rule, that is, a certain rules set is required. Finding solution problem becomes multi-criteria with several alternatives. Alternatives may appear later than the solutions developed.

Scientists and practitioners from all over the world have been searching for optimal solutions to multi-criteria problems. Let's focus on some mechanisms and principles for solving multi-criteria decision-making problems.

Selection by vector criterion, that is, the choice of abstract multiscale extreme selection mechanisms. M.A. Aizerman [1], S.V. Emelyanova [9], I.M. Makarov [19], and B.A. Berezovsky [2] paid attention to this issue.

Under conditions of risk and uncertainty, utility theory is used for multi-criteria selection of alternatives from a discrete set. R.L. Kini [14], H. Reif [14], R.S. Fishburne [10], A.N. Borisov [3] and V.E. Zhukovin [29] addressed this issue in their works.

Application of a multi-criteria selection scheme based on a formalized system of requirements. The system of requirements consists of a set of axioms. The selection scheme is constructed as a consequence of this system. This is mentioned in the works of V. Podinovski [21], E.I. Vilkas [25].

Using convolution of the vector criterion to reduce multi-vector selection to scalar optimization. This mechanism is described in the works of Yu.B. Hermeyer [11], P. S. Krasnoshchekov [17], Yu.G. Yevtushenko [27], R. Stoyer [24], and J.L. Cohon [8].

To solve multi-criteria optimization problems, human-machine procedures are developed and used interactively. This is mentioned in the works of S.V. Yemelyanov [9], A.M. Joffrion [12] and R. Steuer [24].

Application of the Pareto-optimal solution principle for some classes of multi-criteria optimization problems. A domain of compromises is constructed and a set of Pareto-optimal solutions corresponding to this domain is constructed.

These issues were dealt with by Yu.V. Bugaev [5], J. L. Cohon [8], B. Villareal [26], M.H. Karwan [13] and S. Zions [30].

The method of hierarchy analysis developed by T. Saati is used for multi-criteria selection of alternatives [22]. The method is widely used and is firmly established in the theory and practice of multi-criteria selection. To date, several of its modifications and sequels have appeared.

In this paper, we propose one of the directions for the development of the hierarchy analysis method. In this case, it is assumed that the matrix of paired comparisons is the result of some measurement with random errors [16]. Next, estimates of the coefficients of the utility function are calculated, as well as characteristics of alternatives based on statistical methods

We assume that the numbers a_{ij} adequately reflect the expert's preferences about the superiority of the i-th alternative over the j-th in the training sample of m alternatives, and use their values in computational formulas.

Expert behavior models are usually based on the assumption that experts evaluate the parameter with some errors. The expert is considered as a device with metrological characteristics [23]. Based on this principle, we assume that there is a true utility w_i of each i-th alternative and interpret the numbers a_{ij} as realizations of some random variables with mathematical expectations $M[a_{ij}] = w_i/w_j$. We apply the maximum likelihood method (MLM) to obtain statistical estimates for w_i.

The MLM assumes that the combined distribution of random experimental outcomes is known. In this case, there is no such information about the numbers a_{ij}. It is usually assumed that the expert chooses the correct solution (adequate to reality) more often than the wrong one. The frequency distribution of the expert's response (a random variable) monotonically decreases with increasing distance from the distribution center (the true value of the parameter). It is also assumed that the distribution of error values is independent of the size of the object being evaluated.

2   Materials and Methods for Solving Problems Accepted Assumptions

When studying the results of any measurements, including those performed by experts, it is usually assumed that the normal distribution of allowed errors fully corresponds to the above conditions [6]. However, in the case of measuring the ratio of two quantities, the normal distribution is not suitable for reasons of independence from the size of the object. Indeed, it is wrong to assume that, for example, at $w_i = 10$, $w_j = 1$, the errors in estimating the values a_{ij} and a_{ji} have the same order. Under these conditions, the most reasonable assumption seems to be about the logical errors distribution. Therefore, to present MLM, we assume that expert estimation errors are independent and distributed logarithmically normally with the same variances, i.e. the joint density of the distribution of logarithms of a_{ij} values has the form:

$$h(\vec{x}) = \frac{1}{\left(\sigma\sqrt{2\pi}\right)^{m(m-1)}} \exp\left[-\frac{1}{2\sigma^2}\sum_{i,j}\left(x_{ij} - \ln\left(\frac{w_i}{w_j}\right)\right)^2\right].$$

We denote $z_j = \ln(w_j)$, $b_{ij} = \ln(a_{ij})$, form the likelihood function $L(\vec{b})$ and create an algorithm:

$$-\ln(L) = C + m(m-1)\ln(\sigma) + \frac{1}{2\sigma^2}\sum_{i,j}\left(b_{ij} - z_i + z_j\right)^2$$

$$= C + m(m-1)\ln(\sigma) + \frac{1}{2\sigma^2}S(z) \tag{1}$$

where C – constant.

To better understand the results of optimization of the likelihood function, we present the sum S in matrix form. Since S is the sum of the squares of the elements of a certain matrix, we use the obvious identity:

$$\sum_{i,j} d_{ij}^2 = tr(D^T D),$$

which is valid for any matrix D with elements d_{ij}. We introduce a vector z with z_i elements and vector e consisting of m-units (summation vector). Get an idea:

$$S(z) = tr\left\{ \left(ze^T - ez^T - B \right)^T \left(ze^T - ez^T - B \right) \right\}$$

$$= tr\left(ez^T ze^T \right) - tr\left(ez^T ez^T \right) - tr\left(ez^T B \right) - tr\left(ze^T ze^T \right)$$

$$+ tr\left(ze^T ez^T \right) + tr\left(ze^T B \right) + \ldots + tr\left(B^T B \right),$$

where

$$z = \begin{bmatrix} z_1 \\ z_2 \\ z_3 \\ \vdots \\ z_m \end{bmatrix}; e = \begin{bmatrix} 1 \\ 1 \\ 1 \\ \vdots \\ 1 \end{bmatrix}; B = \begin{bmatrix} b_{11} & b_{12} & b_{13} & \cdots & b_{1m} \\ b_{21} & b_{22} & b_{23} & \cdots & b_{2m} \\ b_{31} & b_{32} & b_{33} & \cdots & b_{3m} \\ \vdots & \vdots & \vdots & \vdots & \vdots \\ b_{m1} & b_{m2} & b_{m3} & \cdots & b_{mm} \end{bmatrix}.$$

Let us consider the first term. Since for any matrices P and Q, the identity $tr(PQ) = tr(QP)$ is true, we obtain:

$$tr\left(ez^T ze^T \right) = tr\left[\left(ez^T \right) \left(ze^T \right) \right] = tr\left[\left(ze^T \right) \left(ez^T \right) \right] = m \cdot tr\left(zz^T \right) = m \cdot tr\left(z^T z \right)$$

$$= m \cdot z^T z,$$

since $z^T z$ and $e^T e = m$ are scalars. The rest of the terms are converted in the same way. After the conversion, we get:

$$S(z) = 2z^T \left(m \cdot I - ee^T \right) z + 2z \left(B^T - B \right) e + tr\left(B^T B \right) \tag{2}$$

where I is an identity matrix.

Obviously, w_i optimal values (its estimates using MLM) can be determined with accuracy to an arbitrary multiplier (respectively, z_i - with accuracy to an arbitrary term), so some additional condition like $\varphi(z) = 0$ is needed to normalize the response. In this case, we get a problem with restriction-equality, which can be solved using the Lagrange multiplier method. Let's write down the Lagrange function:

$$\mathcal{L} = m(m-1)\ln(\sigma) + \frac{1}{2\sigma^2} S(z) + \lambda\varphi(z).$$

Now we calculate $\nabla_z \mathcal{L}$ – Lagrange function gradient by the variables z_i. Then:

$$\nabla_z \mathcal{L} = \frac{1}{2\sigma^2} \nabla_z S + \lambda \nabla_z \varphi = 0.$$

As an equality constraint, choose $\varphi \equiv z^T e = 0$. Under the condition $\sigma > 0$, we get the equation:

$$\frac{1}{\sigma^2}\left[2 \cdot \left(m \cdot I - ee^T \right) \cdot z + \left(B^T - B \right) \cdot e \right] + \lambda e = 0$$

or

$$\left(m \cdot I - ee^T\right) \cdot z = \frac{1}{2} \cdot \left(B - B^T - \lambda\sigma^2\right)e.$$

Due to the condition $z^T e = e^T z = 0$, we have:

$$\left(m \cdot I - ee^T\right) \cdot z = m \cdot z - ee^T z = m \cdot z$$

From here:

$$z = \frac{1}{2m} \cdot \left(B - B^T - \lambda\sigma^2\right)e.$$

Let's find λ according to the restriction:

$$\left(e^T z\right) \cdot (2m) = e^T \left(B - B^T - \lambda\sigma^2\right)e = 0.$$

Since the product $e^T(B-B^T)e$ is equal to the sum of the elements of the matrix $(B - B^T)$, and due to its antisymmetry is 0, then:

$$e^T \lambda\sigma^2 e = m\lambda\sigma^2 = 0.$$

Then $\lambda = 0$. We have an estimate:

$$zMLM = 12m \cdot B - BTe. \tag{3}$$

Therefore, w_i will be defined as follows:

$$w_i^{MLM} = \left(\frac{\prod_{j=1}^{m} a_{ij}}{\prod_{k=1}^{m} a_{ki}}\right)^{\frac{1}{2m}}. \tag{4}$$

Calculating the function (1) σ quotient derivative, it is easy to find an estimate using MLM for σ:

$$\left(\sigma^2\right)^{MLM} = \frac{1}{m(m-1)} S\left(z^{MLM}\right). \tag{5}$$

Estimates (3)–(5) were obtained without assuming pairwise comparison matrix inverse symmetry. In other words, it is assumed that the values a_{ij} and a_{ji} are obtained independently of each other and $a_{ij} \neq 1/a_{ji}$. However, in practice, the expert usually fills only the matrix A upper part, and for the lower half, $a_{ij} = 1/a_{ji}$ is assumed. In this case, matrix B will be antisymmetric and function (1) will take the form:

$$-\ln(L) = C + \frac{m(m-1)}{2} \ln(\sigma) + \frac{1}{2\sigma^2} \sum_{i<j} (b_{ij} - z_i + z_j)^2$$

$$= C + \frac{m(m-1)}{2} \ln(\sigma) + \frac{1}{2\sigma^2} S_1(z). \tag{6}$$

Accordingly, we will get estimates:

$$z^{MLM} = \frac{1}{m}Be;$$

$$w_i^{MLM} = \left(\prod\nolimits_{j=1}^{m} a_{ij}\right)^{\frac{1}{m}};$$ (7)

$$\left(\sigma^2\right)^{MLM} = \frac{2}{m(m-1)}S_1\left(z^{MLM}\right).$$ (8)

The values (8) are used in the hierarchy analysis method as approximate formulas for calculating the components of the eigenvector of the matrix A, which are sufficiently accurate for small m. In this case, we obtained them as maximum likelihood estimates for components of the same vector, provided the lognormal distribution of expert estimation error. Expression (8) gives σ^2 parameter estimate, which characterizes the variance of errors in expert evaluation of matrix a elements. Obviously, in the case of a consistent matrix, A will be $S_1 = 0$, and, consequently, $\sigma^2 = 0$.

So, we found estimates not only of the utility of alternatives, but also of the variance of expert errors in the case of obtaining an inconsistent matrix of paired comparisons.

In a similar way, you can find estimates using the maximum likelihood method for the coefficients of the generalized criterion function of a given structure. Let the vector x^r correspond to r-alternatives and let us have generalized criterion function $F(x, b)$, where b is coefficients vector for evaluation. Imposing constraints $F(x^r, b) = w_r$ and determining the corresponding conditional maximum L point, we can obtain coefficients b_u using MLM estimates.

This estimate can be obtained by considering not all the many alternatives, but a training sample with subsequent extrapolation of the found function of the generalized criterion to the entire set of alternatives in order to order it. In other words, the proposed method for processing the matrix of paired comparisons can be considered as a new modernized version of the method of extrapolation of expert estimates. In the case where generic criterion function is:

$$F(x) = \exp\left(\sum b_u f_u(x)\right) = \exp\left(b^T f(x)\right),$$ (9)

where $f_u(x)$ – are some known functions. $u = 1,...,k$, you can get an explicit formula for maximum likelihood estimates of b_u coefficients. Let Φ be a matrix with the elements $\Phi_{ru} = f_u(x^r)$. Then, in the expression (2), $z = \Phi b$. However, the normalization condition $z^T e = 0$ is required.

Denote $W = I - ee^T/m$. It is easy to make sure that matrix W has idempotency property, i.e. $W^T = W$ and $W^2 = W$. It is also easy to show that for any vector p the vector Wp has zero component sum. This means that multiplying the matrix W by any vector "averages" it, this is equal to $Wp = p - \bar{p}e$. Then the normalization condition will be automatically satisfied if you use the $z = W\Phi b$ representation. Let's substitute this expression in (2). We have:

$$S_1 = mb^T \Phi^T W^3 \Phi b + b^T \Phi^T W\left(B^T - B\right)e + \frac{1}{2}tr\left(B^T B\right).$$

Therefore, for maximum likelihood estimation, there must be

$$\nabla S_1 = 2m\Phi^T W\Phi b + \Phi^T W\left(B^T - B\right)e = 0 \tag{10}$$

due to equality $W^3 = W$.

Let us assume that $k \times k$-size matrix $\Phi^T W\Phi$ also has a full rank, i.e. it is not degenerate. In this case, by (10), we have a formula for estimates according to generalized criterion coefficients MLM:

$$b^{MLM} = \frac{1}{2m} \cdot \left(\Phi^T W\Phi\right)^{-1}\Phi^T W\left(B - B^T\right)e. \tag{11}$$

Generalized criterion coefficients assessment can be received as well by pair comparisons matrix processing traditional method means accepted in hierarchies analysis method [16]. To do this, you need to obtain the matrix A eigenvector w, and then select generalized criterion coefficients that satisfy condition:

$$F\left(x^i\right) = w_i. \tag{12}$$

If coefficients number is less than alternatives number, then Eqs. (12) are likely to be performed only approximately, so the proximity condition $F(x^i)$ and w question arises. For example, this may be the usual least squares principle, but most likely, a special method is needed for this approach. In any case, this approach is indirect: two, in general, unrelated problems are solved - the search for utility estimates and approximating function selection [4, 5, 7]. The proposed method in the present work is direct, since coefficients selection is carried out on the generalized criterion values proximity basis to real (and not approximated through an eigenvector) expert comparisons results. Therefore, with poor paired comparisons matrix elements consistency, one can expect greater accuracy from the proposed method.

3 Results

The rating task refers to multi-criteria decision tasks [20].

Let's consider the computer implementation of the modification of the method of extrapolation of estimates for the analysis of performance indicators of educational organizations. Performance is evaluated using seven main indicators and one additional indicator. All selected key figures are in turn decomposed into smaller ones. For each indicator, values are mapped, which mainly differ between educational organizations. In particular, these values may be the same.

It is proposed the following algorithm to prepare data for processing by the method of extrapolation of expert estimations [4–6].

Key indicators are:

– educational activities;
– scientific activities;
– international activities;

– financial and economic activities;
– infrastructure;
– employment and staffing.

We will consider them as global indicators, and the indicators that they consist of are local.

Among all analyzed educational organizations the maximum and minimum values are allocated, after normalization is carried out for the maximum according to the formula (13) for each local indicator:

$$f'(x) = \frac{f(x) - f_{min}(x)}{f_{max}(x) - f_{min}(x)}, \tag{13}$$

where $f(x)$ – current indicator value; $f_{min}(x)$ – indicator minimum value among educational organizations, $f_{max}(x)$ – indicator maximum value among educational organizations. The better the score (its value is greater), the higher the efficiency.

Then all local values are summed up within each global criterion. The obtained value is assigned to the global criterion. Since local criteria number in each global criterion is different, this can lead to global criteria obtained values heterogeneity. Therefore, we repeat the normalization for each global criterion.

As described actions result, we obtain alternatives set (educational institutions) with homogeneous seven main evaluation criteria values. After that, you need to select a training sample, and then start comparing using expert assessments extrapolation method.

Based on the mathematical model the program algorithm and the software product structure were developed. In the next step, the information system user interface was created. Alternatives automatic selection based on bisection algorithm eliminates preferences non-transitivity [7]. In other modes there may be inconsistency in the making decision person preferences [15, 18].

Proposed in the paper preferences extrapolating expert assessments method was implemented in executable module *Project.exe* form. The program is implemented in the MS Windows 8.1 environment for IBM-compatible computers in the object-oriented language "C++." MS Visual Studio has been selected as the executable development tool.

The developed application package is not focused on a specific subject area and in this sense is universal [15, 28].

The software application includes the following five main function modules:

1. *Unit1* – the main module, in which additional information is entered to solve the problem and the existing one is connected or a new database is created, the main criteria are selected, and its normalization method is indicated;
2. *Unit2* is a module for editing existing databases;
3. *Unit3* is a module for filling indicators numerical values for each database alternative;
4. *Unit4* - the module in which the examination is carried out on an ordinal scale;
5. *Unit5* - reverse criterion function coefficients numerical values are calculated, final alternatives ranking is carried out.

Communication between modules is carried out through formal and actual parameters lists. When you start the program for the first time, you must populate the database for the selection task to be solved.

When forming a sample, data from database tables are entered into the intermediate data array $A = \{a_{ij}\}$, where $i = 1,...,M$, M - analyzed educational organizations number, $j = 1,...,N$, N - educational organizations evaluation indicators number. Sampling is made from all tables according to the selected educational organization. The values are then normalized for each key figure and entered into a new intermediate array of normalized values. This data is then transmitted for comparison.

The normalization inevitability is that there is a fairly large variation in indicators numerical values. In this case, there are great difficulties in specifying the indistinguishability threshold. This is inconvenient in computational procedure terms.

If it is necessary, you can make changes to the existing database. Then you can specify a value for the indistinguishability threshold at which the criteria will be considered the same. It is recommended that you select a threshold value not more than 5% from the largest criteria value in the numeric equivalent.

All necessary data should be included in possible alternatives comparison. Alternatives ranking occurs on an ordinal scale. Moreover, there are three modes for selecting pairs of alternatives for comparison: manual, automatic, and all possible comparisons. In the first case, the choice of a pair of compared alternatives is made by the user or expert.

From *Unit3* module, the populated database is transferred to *Unit4* module. In this module, alternatives pairwise presentations for examination are formed on the obtained database basis on an ordinal scale. As a result of such comparisons, a matrix containing difference numerical values between compared alternatives corresponding criteria is formed. Table 1 is formed from this data.

Table 1. Structure of the matrix of paired comparisons.

Presentation number	$-\xi 1$	$-\xi 2$	...	$-\xi n$
1	$f_i(x_1) - f_j(x_1)$	$f_i(x_2) - f_j(x_2)$	...	$f_i(x_n) - f_j(x_n)$
2	$f_l(x_1) - f_t(x_1)$	$f_l(x_2) - f_t(x_2)$	...	$f_l(x_n) - f_t(x_n)$
...	...	...	...	...
P	$f_p(x_1) - f_s(x_1)$	$f_p(x_2) - f_s(x_2)$	...	$f_p(x_n) - f_s(x_n)$

A system of inequalities is formed based on the data in Table 1. Next, the basic points of this system are found for detecting and eliminating non-transitivity of expert preferences. If the expert wishes to interrupt the examination, the above matrix is passed to the Unit5 module. In this module determined the numerical values of the coefficients of functions of generalized criterion, compute values of this criterion for each alternative and a final ranking of alternatives in descending order of the function values of the generalized criterion. The results of this stage are entered in Table 2. At the user's request, the information can be printed (Fig. 1).

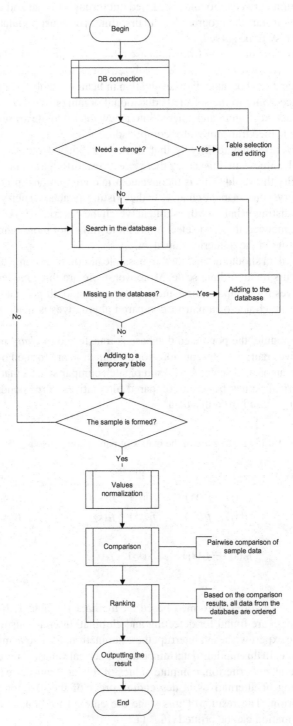

Fig. 1. Program flow chart.

Table 2. The result presentation.

Alternative	General criterion function value
1	$F(1)$
...	...
M	$F(M)$

The program starts from the main window. In order for the software module to work successfully, you must connect database with completed performance tables. You can do this using *File -> ConnectDatabase*. For this example, the database consists of eight tables. The tables contain the following information:

- organization full name, location and address (ORG);
- main activity indicators (MA);
- research activity indicators (RA);
- international activity indicators (IA);
- infrastructure available to the organization indicators (INF);
- financial and economic activity indicators (FEA);
- etc.

After database connection you can either start the organizations rating evaluation, or edit database (Fig. 2).

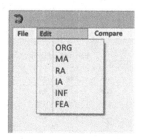

Fig. 2. Menu for selecting database tables to edit.

You can open database tables for editing using *File -> OpenTable* menu. This opens table selection window shown in Fig. 3.

The Fig. 4 shows a screen form for analyzing the effectiveness of educational organizations.

It consists of three tabs. *Sample* tab searches for an educational organization in database, adds one or more organizations to the selection list using *Select* button. If the list is generated, you need to click on the appropriate button. The next tab compares educational organizations against calculated standardized indicators (Fig. 5).

Fig. 3. The selection window of the database table.

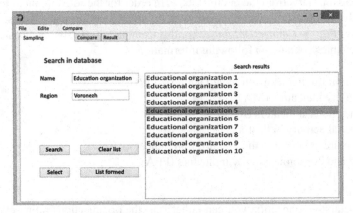

Fig. 4. Sample generation window.

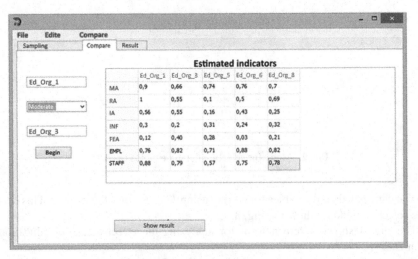

Fig. 5. Alternative comparison window.

Based on comparison results, recovery function coefficients are calculated according to (11). Then, normalized values of performance indicators are calculated for all educational organizations from the database, which are multiplied by the calculated coefficients of the utility function. As a result, we get a numerical value for each organization. These values can rank organizations and determine their rating. The results of rating calculation are shown in Fig. 6.

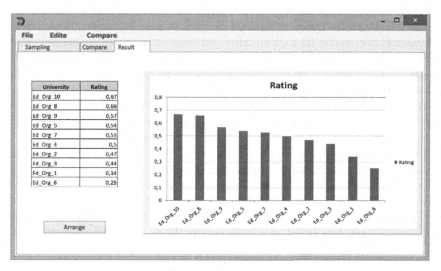

Fig. 6. Rating results.

4   Conclusion

Monitoring for a long time will be the main data source on n higher education. Its data are relatively reliable, although there is room for improvement. One possible step that would help to assess data validity indicators was to compare the monitoring information with information on the same variables from unknown sources. Unfortunately, there were no such sources for most variables. However, public activity data aggregated directly by educational organizations from the Russian Scientific Citation Index (RSCI), Scopus and Web of Science databases are available for verification. In general, it should be noted that for both research and management purposes it is desirable to use indicators against which it is possible to verify information accuracy submitted using independent sources. The general knowledge that information could be verified was likely to improve its presentation quality.

Data reliability considerations are important; however, they are not the main obstacle to monitoring data use in higher education research. The most variables distributions specificity significantly limits investigator. For almost any monitoring data analysis, the main recommendation will be robust emission-resistant analysis methods using. Another possible solution is data vindication and variables transformation. For variables number with low variability, binarization is the best option.

Another limitation is the high indicators correlation. Therefore, a single indicator is necessary. Using variables together only complicates models, but does not improve them in predictor power terms. It also leads to multicollinearity and, possibly, instead of noticeable differences in the population, differences will be found in individual cases that are exceptions. A "hint" for selection can be variable distribution (the optimal distribution is far from all indicators). For some analysis forms, a possible strategy is to exclude cases that do not matter by the corresponding variable. But this can lead to a shift in the conclusions obtained when excluding such educational organizations.

The initial data for the computational experiment were educational organizations additive indicators values for each criterion, and additive indicators normalized values.

As a result of the computational experiment, the function of the generalized criterion was obtained. The application of the developed software product gave a ranking of the analyzed educational organizations according to the effectiveness of their activities.

Thus, the paper describes data preliminary preparation for processing using extrapolating expert estimates method. Tools for selecting alternatives in the form of the Project software complex invariant to the subject area have been built.

Tools for selecting alternatives in the form of the "Project" software package, which is invariant to the subject area, are constructed. Computational experiments have been performed that confirm the possibility of using the developed models, algorithm, and individual choice mechanism to solve problems of finding the best alternative in the case of several criteria. A software product has been developed that implements models and selection algorithms based on new variants of the expert evaluation extrapolation method.

References

1. Aizerman, M.A., Aleskerov, F.T.: Choice of Options: Fundamentals of Theory. Nauka, Moscow (1990)
2. Berezovsky, B.A., Baryshnikov, Yu.M, Borzenko, V.I.: Multi-criteria Optimization: Mathematical Aspects. Nauka, Moscow (1989)
3. Borisov, A.N., Vilyums, E.R., Sukur, L.Ya.: Dialog Decision-Making Systems Based on mi-Ni-Computer: Information, Mathematical and Software. Zinatne, Riga (1986)
4. Bugaev, Yu.V., Korobova, L.A., Shurupova, I.Yu.: Search for all solutions to the problem of dynamic programming in the case of coincidence of their multi-criteria estimates. J. Bull. Voronezh State Univ. Eng. Technol. 82(1), 398–403 (2020)
5. Bugaev, Yu.V., Nikitin, B.E., Chernyaeva, S.N.: Narrowing the Pareto set using expert estimates. J. VSU Bull. Ser. Phys. Math. 2, 160–164 (2006)
6. Bugaev, Yu.V., Nikitin, B.E., Chernyaeva, S.N.: Processing of the matrix of paired comparisons by the maximum likelihood method. J. Control Syst. Inf. Technol. 3(33), 35–38 (2008)
7. Bukharin, S.V., Melnikov, A.V., Chernyaeva, S.N., Korobova, L.A.: The method of immersion the problem of comparing technical objects in an expert shell in the class of artificial intelligence algorithms. IOP Conf. Ser. Mater. Sci. Eng. (2017). International Conference on Materials, Article no. 012208
8. Cohon, J.L.: Multiobjective Programming and Planning. Academic Press, New York & London (1978)

9. Emelyanov, S.V., Larichev, O.I.: Multi - Criteria Methods of Decision-Making. Znanie, Moscow (1985)
10. Fishburne, P.S.: Theory of Utility for Decision. Nauka, Moscow (1978)
11. Hermeyer, Yu.B.: Introduction to the Theory of Operations Research. Nauka, Moscow (1971)
12. Joffrion, A., Dyer, J., Feinbere, L.: Solving optimization problems under many criteria based on human-machine procedures. J. Issues Anal. Decis. – Making Proced. 126–145 (1976)
13. Karwan, M.N.: On finding starting solutions for some specially structured linear programming problems. Working Paper 445, School of Manadgment, State University of New York, Buffalo (1980)
14. Keeney, R.L., Rife, H.: Decision-Making at Many Criteria: Preferences and Substitution. Radio and Communications, Moscow (1981)
15. Korobova, L.A., Savvina, E.A., Kovaleva, E.N., et al.: Application of cluster analysis for business processes in the implementation of integrated economic and management systems. Advances in Economics, Business and Management Research. In: Proceedings of the Russian Conference on Digital Economy and Knowledge Management (RuDEcK 2020), pp. 316–323 (2020)
16. Kou, G., Chen, Y., Ergu, D.: Pairwise comparison matrix in multiple criteria decision making. J. Technol. Econ. Dev. Econ. **22**(5), 738–765 (2016)
17. Krasnoshchekov, P.S.: Mathematical Models in Operations Research. Nauka, Moscow (1984)
18. Kuo, T.: Interval multiplicative pairwise comparison matrix: consistency, indeterminacy and normality. J. Inf. Sci. **517**, 244–253 (2020)
19. Makarov, I.M., Vinogradskaya, T.M., Rubchinsky, A.A.: Theory of Choice and Decision – Making. Nauka, Moscow (1982)
20. Nikitin, B.E., Ivliev, M.N., Korobova, L.A.: Calculation and analysis of the rating of scientific periodicals. J. Bull. Voronezh State Univ. Eng. Technol. **79**(4), 97–103 (2017)
21. Podinovsky, V.V., Nogin, V.D.: Pareto-Optimal Solutions to Multi-criteria Problems. Nauka, Moscow (1982)
22. Saaty, T.L.: Multicriteria Decision Making. The Analytic Hierarchy Process: Planning, Priority Setting, Resource Allocation. University of Pittsburgh, RWS publications, Pittsburgh (1990)
23. Shmetterer, L.: Introduction to Mathematical Statistics. Nauka, Moscow (1976)
24. Steuer, R.M.: Multicriteria Optimization: Theory, Calculations and Applications. Radio and Communication, Moscow (1992)
25. Vilkas, E.Y., Maimenas, E.Z.: Solutions: Theory, Information, Modeling. Radio and Communication, Moscow (1981)
26. Villareal, B.A., Karwan, M.N., Zoints, S.: Branch and bound approach to interactive multicriteria integer linear programming. J. Paper presented at Joint National Meeting TIMS/ORSA: Washington, D. C. (1980)
27. Yevtushenko, YuG, Potapov, M.A.: Numerical methods for solving multicriteria problems. J. Cybern. High Technol. **3**, 209–218 (1987)
28. Zhou, X., Deng, Y., Hu, Y., Chan, F.T.S., Ishizaka, A.: A DEMATEL-based completion method for incomplete pairwise comparison matrix in AHP. J. Ann. Oper. Res. **271**(2), 1045–1066 (2018). https://doi.org/10.1007/s10479-018-2769-3
29. Zhukov, V.S.: Models and Decision-Making Procedures. Manerba, Tbilisi (1981)
30. Zions, S.: Multiple criteria decision making for discrete alternatives with ordinal criteria. J. Working Paper 299, School of Manadgment. State University of New York, New York, Buffalo (1977)

Information Retrieval Approach Using Semiotic Models Based on Multi-layered Semantic Graphs

Alena Korney[1]([envelope]) [iD], Elena Kryuchkova[1] [iD], and Vitaly Savchenko[2] [iD]

[1] Polzunov Altai State Technical University, Lenin Avenue 46, 656038 Barnaul, Russia
korney.alena@yandex.ru, kruchkova_elena@mail.ru
[2] E-System LLC, Depovskaya Street 22, 656015 Barnaul, Russia
64svv@rambler.ru

Abstract. The paper presents a novel method of information retrieval using semiotic models based on multilayer semantic graphs. The layers correspond to the level of detail of the extracted semantic information. The key developments underlying the method are described. Firstly, it is a graph model, which can be automatically built on the basis of general linguistic dictionaries or specialized encyclopedias. Using formalized text sources allows us to build a graph in a reasonable amount of time. The graph vertices are presented by canonical forms of words. Three types of connections are used as graph edges: association, definition, and synonymy. A method of automatic selection of a connection type based on the analysis of the dictionary entry structure is proposed. Secondly, an approach to using the semantic graph as a flexible basis for complex information retrieval systems is described. Information about existing connections between lexical units can be interpreted in different ways. Experiments show that, depending on the task, information about synonymous or associative relationships may come to the fore. In general, data extracted from dictionaries are reliable and complete enough to use the presented graph as a framework for information retrieval systems.

Keywords: Semantic graph · Information retrieval · Information extraction · Image classification · Semantic search

1 Introduction

1.1 Background

The exponential growth of information has made many tasks inaccessible to manual execution. Let's consider such tasks as image recognition, understanding natural text (such as meaning extraction, or categorization). These tasks have a lot in common in terms of human perception, and their solution usually boils down to the extraction of some known data. The success achieved over the past decade in the field of computer vision, information retrieval, sentiment analysis, etc., looks optimistic. But a few years ago, this success was mainly achieved through the use of highly specialized techniques.

© Springer Nature Switzerland AG 2020
V. Jordan et al. (Eds.): HPCST 2020, CCIS 1304, pp. 162–177, 2020.
https://doi.org/10.1007/978-3-030-66895-2_11

Modern studies, on the other hand, demonstrate that significant quality improvements can be achieved by combining different approaches from different areas.

One of the best examples of a technique combining is the "Bag of Features" approach used for image classification. The task of image classification has many variations - from the simplest single tag assigning to complex multiclass classification with the object localization. Basic approaches have traditionally used feature extraction in conjunction with well-known classifiers, such as the Naive Bayesian method, the nearest neighbor method, SVM, etc. There is also a group of approaches based on the Bag of Words model, borrowed from natural language processing [1]. These approaches consider the image as a document with a special dictionary, and each visual feature is used as a word with its own meaning. This is a very important step towards high-level information processing. The BoW model is not the only way to incorporate semantics into image processing. Word Embeddings approach [2–4] is one of the most popular methods of language modeling and feature learning. Each word can be represented by a vector, and its value depends on the word frequency and word co-occurrence distribution. Papers [5, 6] show how this idea can be used for multiclass image classification.

The multiclass classification task is related not only to visual data. Some textual information should also be analyzed in this way (e.g. user reviews, movie/hotel reviews, etc., where different functions and aspects can be described and evaluated differently). These tasks depend on the domain and require semantic models (e.g., word embeddings [7, 8]) and reliable sources of general and context-dependent knowledge [9]. SentiWord-Net [10], SenticNet [11], WordNet Affect [12] are examples of semantic resources in which heterogeneous features can be used together: semantic relationships are combined with sentiment marks, high-level concepts, and contextual information.

It is becoming obvious that semantic models and complex interdisciplinary approaches are now in the trend. Unfortunately, models and resources described above have more than just strengths. They also have some drawbacks. For example, word embeddings do not allow different types of semantic relationships (synonyms, definitions, etc.). The statistics are not flexible enough. Word disambiguation is not included by default in most cases, and then requires some workarounds. Semantic resources such as WordNet Affect, SenticNet, and SentiWordNet are more flexible for some tasks. They allow you to use heterogeneous semantic relationships, store and extract contextual data. But these resources are widely used, and their structure is often redundant for specific, specialized tasks.

1.2 Aim

We propose to consider different recognition systems as samples of the general information retrieval task based on the corresponding semiotic model with a semantic graph as a core. The key idea was to combine certain concepts, words, and connections into one system. This system can be considered as a particular implementation of the general relationship that exists in the main graph. It is obvious that automated search systems need a reliable source of some general knowledge to simulate the human-like analysis process.

The main purpose of this work was to build a semantic graph of the Russian language, which would have the following important properties:

1. Semantic content must be domain-independent.
2. The graph structure must be simple and flexible.
3. Automated building and extension must be available.
4. Semantic content must be full, reliable, and adequate.

The domain independence property, which can be considered the most important, deserves special attention. There is a natural question - what does domain independence mean? Of course, if we talk about everyday life and the world around us with all its diversity, domain independence is obligatory. But if we talk about a specific scientific field, its dictionary has specific terminology. For example, the lexicon of a mathematical encyclopedia overlaps poorly with the common lexicon. This intersection is usually non-terminological or has a different meaning. For example, "root" has nothing to do with plants, "ring" doesn't mean the object of the appropriate shape, "cortege" isn't related to transport, and "operator" is not about the profession. Therefore, such an intersection is unsuitable for the analysis of documents of a highly specialized area.

Thus, let's consider the domain independence as independence from the specific features of a large number of subdomains (e.g., common lexicon, mathematical, medical, etc.).

The requirements of automatic build and semantic content fullness are sufficiently provided by using explanatory dictionaries of the chosen language. But these dictionaries are not suitable for building graphs for specific scientific domains.

The graphs described in this paper were designed to overcome the shortcomings of existing models. The key goal was to create an easy, flexible, and expandable model and to provide reliable and accurate data based on a well-known source of common-sense knowledge.

Section 2.2 contains statistical analysis results of data extracted from selected dictionaries.

Section 2.3 proposes a group of approaches based on a semantic graph developed and some practical results achieved using these approaches for image classification and semantic search.

2 Basic Semantic Graph

The basic semantic graph is implemented based on of general linguistic dictionaries of the Russian language. The structure of the lexicon described below is an extension of the mathematical model presented in [13, 14]. Two dictionaries were chosen as the source of semantic data:

- Explanatory dictionary of the Russian language by Ozhegov and Shvedova [15].
- Dictionary of Russian synonyms and words with close meanings [16].

Dictionaries were automatically analyzed using RML[1] semantic parser.

---

[1] http://aot.ru.

2.1 Mathematical Model of Multi-layered Semantic Graph

Before defining the multi-layered semantic graph model used in this paper, let's look at some of the most important concepts and relationships between them. Let A be the alphabet of the natural language. In this work, the Russian language was selected for practical implementation and experiments.

Let the finite set of concepts $V \subset A^+$ be a lexicon of a model. Members of V are used as vertices for the semantic graph. In common practice, each concept can be both a single word and a collocation [17]. Moreover, elements (words) of a lexicon can denote not only objects but also events, features, processes, etc. It means that corresponding vertices can be presented in a graph.

The model considers morphological features of individual words as unimportant. It means that words are presented in V by their canonical forms. Thus, both in the construction of the semantic graph and in the process of its use, the text is pre-processed by stop-word removal and stemming or lemmatization. Edges of the graph connecting particular concepts correspond to the type of relationship between them.

During cognitive development, human builds a lot of complex associative connections, containing a whole range of related properties and characteristics. These connections and concepts become a part of a known world model. The use of this related knowledge allows us to understand and refine the incoming information.

In human recognition of the large text main theme, the associative connections between the concepts encountered in the text play a key role. If several semantic concepts of the same domain appear in the text fragment being analyzed, we can say with great confidence that this fragment relates to the relevant domain.

Obviously, the association is one of the most fundamental relationships between concepts. But synonymy and definition relationships are equally important for text understanding because we can use the knowledge about the basic underlying object and transfer them to a defined object.

Thus, we should consider all these types of semantic relations. Each relation is presented as a tag associated with a graph edge, and all tags are combined into the set of three elements:

$$L = \{l_a, l_s, l_d\}, \tag{1}$$

where l_a is association relation, l_s is synonymy relation, and l_d is definition relation.

These types of relationships will be automatically extracted from existing dictionaries. Generalized world knowledge can be extracted with sufficient confidence from intelligent dictionaries of natural language, dictionaries of synonyms, etc. General linguistic dictionaries describe interconnected objects, events, phenomena of a single and indivisible world around us. We hold the opinion that the corresponding semantics of interconnections should be stored in only one semantic graph of the basic (or zero) level.

The need for additional layers arises when moving to highly specialized areas, which will be considered further. In general, the semantic graph model contains $n \geq 0$ number of layers. Let $G_i = (V_i, U_i)$ be a graph of the i-th layer. Graph G_0 corresponds to the basic layer filled with data extracted from common dictionaries. An each G_i graph at $1 \leq i \leq n$ corresponds to the i-th layer of semantic expansion.

All vertices $V_i \subset V$ of all layers are marked with words from V. All the edges in U_i are tagged with relationship label L.

Thus, the model of the multilayer semantic graph used in this work is described as a tuple:

$$M = (A, V, L, n, G_0, G_1, \ldots, G_n). \tag{2}$$

Let's consider ways of extracting relations from linguistic dictionaries.

Synonymy Relation. Automatic extraction of synonymous relationships is a simple task with low computational complexity due to the presence of special dictionaries synonyms. Besides, explanatory and encyclopedic dictionaries usually also have information about synonyms presented by a simple template: $<concept>$-$<equivalent\ concept>$.

The main problem is related to l_s edge direction in a graph. In the case of explanatory dictionary entry analysis, we definitely can say that each equivalent concept reference means bidirectional synonymy. But it's not true for the dictionary of synonyms and words with close meanings. Therefore, a one-directional connection l_s is built from a defined word to each word of the list of its synonyms in the dictionary.

Definition Relation. Definition relation l_d is the most complex task for automated extraction. Generally speaking, the syntax design of dictionary entries can be very different. The defined concept is not always highlighted by special signs, so a way to detect it is necessary.

The method chosen by the authors showed good results in the processing of the dictionary. The method is based on the search for the main word w of dictionary entry $\phi = \alpha_1 w \alpha_2$. The word w is the main word of phrase ϕ if w does not appear as a dependent word in any combination from ϕ. If the phrase contains more than one satisfying word, we choose the first of them as the main.

The word w_{def} is a definition of word w if w_{def} is the main word of the corresponding explanatory dictionary entry ϕ. In most cases, definition relations connect the word with its hyponym.

Association Relation. Association relation l_a is built as one-directional edge between two words w_1 and w_2, if w_2 appears in corresponding explanatory dictionary entry.

Strictly speaking, all the relation selected for automatic extraction are fuzzy relations of type $V \times V \rightarrow [1..0]$, and the weight of each relation corresponds to the system's degree of confidence in the fact that relation of a particular type exists for connected words.

Semantic Proximity of Words. Let $G_i = (V, U_i)$ be the oriented semantic graph of i-th level, where V is a set of words, U_i is a set of edges corresponding relation types from L.

The farther apart in the G_i graph are the vertices (words or corresponding objects), the less they are semantically related. And therefore, it's less likely to get them into one semantic cluster. It means that semantic relatedness of main topics of texts containing unrelated words is extremely unlikely.

Let semantic relatedness probability $p(x, y)$ for directly connected words x and y be equal to the weight of the corresponding edge u$(x, y) \in U_i$. Considering the weight of the path as a probability of a joint event, we will find that the weight of the path does not exceed $max(p(v_i, v_{i+1}))^n$, where $x = w_1, w_2, \ldots w_n = y$ – a path from x to y, and n is the length of this path.

Let's determine the measure of proximity of two objects as the weight of the path between them. Considering the semantic proximity of words as the maximum probability of a joint event, we choose the path of $R_j(x, y)$ with the maximum corresponding value:

$$d(x, y) = max_{r_j}\left(\prod_{(w_{k-1}, w_k) \in R_j} p(w_{k-1}, w_k)\right). \tag{3}$$

Considering the weight of the edge as a relationship between concepts with some degree of confidence, we choose the weight $p(w_{k-1}, w_k) < 1$.

Consequently, the expression (3) value quickly becomes smaller than some of the specified ε. Moreover, to increase the influence of the length of the path between objects, it is necessary to enter the $\gamma < 1$ damping factor, in which the value of $d(x, y)$ fades even faster:

$$d(x, y) = max_{r_j}\left(\prod_{(w_{k-1}, w_k) \in R_j} p(w_{k-1}, w_k) * \gamma^k\right). \tag{4}$$

Thus, the neighborhood $O(x, \varepsilon)$ can be defined as a set of objects $y_i \in V$ with proximity value $d(x, y_i) > \varepsilon$. Obviously, the structure and contents of the neighborhood depend significantly on the weights of the relationship presented in the graph. Since the structure and weight of the relationship in the graph are conditioned by semantic connections, the neighborhood built based on G_i we call *the semantic neighborhood*.

Figure 1 shows the contents of the semantic neighborhood of the word "bus" in the graph G_0 (basic level).

2.2 Analysis of Semantic Data Extracted from Basic Dictionaries

During the automated parsing process, ≈ 80000 dictionary entries were analyzed. The majority of them (≈ 61000) were extracted from the explanatory dictionary. The basic graph contains 44771 words and ≈ 200000 relations. Table 1 represents part of speech statistics. Table 2 represents relation statistics for connected word pairs.

The majority of extracted word-to-word semantic connections are presented by "clear" relation (only one type of relation exists for word pair). But there are some words connected with 2 or 3 types of relations. This fact may be interpreted as a "strong" connection on the one hand. But on the other hand, we should remember such features of natural language as polysemy, homonymy, etc.

2.3 Practical Usage of Basic-Level Semantic Graph

Data extracted from common dictionaries are domain-independent, reliable, and full enough to represent some basic world knowledge. These features allow using the given semantic graph as a framework for complex information retrieval systems. Information about existing word-to-word connection can be interpreted differently: in some tasks,

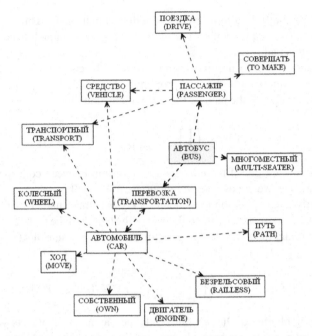

Fig. 1. Semantic neighborhood of the word "bus" in the graph G_0.

Table 1. Part of speech statistics.

Part of speech	Canonical forms count
Noun	23298
Adjective	7830
Participle	295
Infinitive	12106
Verb	74
Adverb	1168

association means more than definition and synonymy; in other tasks, synonymy is the best source of semantic information. We demonstrate this fact by the next experiments carried out by the authors.

Complex Image Classification. Image classification is a good example of how you can use the associative connection presented on the graph. Everyone uses some knowledge about the world to interpret visual information. For example, we can recognize a large and complex object by its part, shadow, or simplified form. Parts of the object are strongly semantically related to each other or the high-level concept (parent object). This feature can significantly improve the quality of the classification of images.

Table 2. Relation type statistics.

Synonymy	Association	Definition	Count of edges with combination of types
+	−	−	32875
−	+	−	111449
−	−	+	49949
+	+	−	638
+	−	+	2306
−	+	+	1507
+	+	+	102

Because of the nature of visual information and its discrete representation methods, the image classification task is considered a challenge. Classification approaches are usually based on the presentation of images as a set of visual words, and then converting them into a histogram of appearance frequencies. However, the use of a set of separated classifiers has some drawbacks. Information on natural connections between certain classes or between the inner parts of an individual class is not used.

The paper [18] explores the possibility of using the information in a semantic graph to classify images. The semantic graph was expanded by stable phrases with a stable visual representation. Phrases injected into the graph are strongly semantically associated with their main words. This method has improved the relevance of classification results through semantic correction. Complex images that contain multiple instances of related classes can be successfully categorized using this method.

Semantic Search in Large Text Collections. Semantic search in large text collections is a very difficult task. The complexity of this task is due to several reasons, such as the size of the collection, the quality and performance of the evaluation of semantic relevance, the size of the request, etc. We have to deal with such tasks as synonym replacement processing, the use of common concepts (for example, when the text about the dog uses "pet" etc.).

The work [19] is devoted to novel semantic search approach, used for multi-thread analysis of large texts. A key feature of the proposed system is the lifting of restrictions on the size of the search query. This paper proposes the development of the method, conducted experiments on the "noise" of the text, demonstrating the quality of analysis.

The source data are presented by text collection and the search query. The main assumption is that typical text collection is heterogeneous. It means that some fragments are more informative for the semantic comparison process than others. So, each collection can be split to the fragments (pages, paragraphs, sets of sentences, etc.) called windows.

Let $G_{sem} = (V, U)$ be a semantic graph of text fragment. G_{sem} is a directed graph where V is the set of vertices (each vertex is presented by a canonical form of Russian word) and U is the set of edges and each edge is defined by some semantic relation type and the weight. The edge direction depends on the relation type. We define two edges

u_α and u_β as equivalent if they connect equivalent words by the same relation type in the same direction.

Each sentence of the window is being analyzed using RML semantic parser. The semantic graphs of particular sentences are used to build the semantic graph of the whole window. The process of window graph building includes three stages: iterative addition of all sentence graphs, normalization, and threshold-based cut-off.

The resulting graphs must be compared to evaluate the relevance of the search query and the current window. Relevance evaluation based on the maximum common edge subgraph is not suitable for this task because of the following reasons:

- The maximum common edge subgraph problem on general graphs is NP-complete;
- Text style can influent semantic graph structure, so two texts with the same meaning can have different graphs (and low value of relevance).

The core idea of the presented system is using a semantic graph of the Russian language as a source of common knowledge. Using a dictionary-based graph allows us to find some common concepts for different words presented in the search query and target window and then use these concepts to evaluate relevance degree.

Let $P(a_k, a_m)$ be a relatedness factor of two words a_k and a_m from request graph $G_{request}$ and window graph G_{sem} respectively. $P(a_k, a_m) = 1$, if a_k is the same word as a_m, and $P(a_k, a_m) \in [0..1)$, if it's not. Let $i_0 i_1 \ldots i_q$ be the path from a_k to t and $j_0 j_1 \ldots j_n$ be the path from a_m to t, where t is the closest common ancestor for a_k and a_m. $P(a_k, a_m)$ may be obtained as follows:

$$P(a_k, a_m) = P(a_m, a_k) = \prod_{i=1}^{q} p_{i-1,i} * \prod_{j=1}^{n} p_{j-1,j}, \qquad (5)$$

where $p_{i-1,i}$ – a weight of the edge between i-1 and i; $p_{j-1,j}$ – a weight of the edge between $j - 1$ and j vertices.

The final stage of relevance evaluation is the selection of the best match of search request and window. Each match is evaluated by the relevance factor S (maximum value of S corresponds to best match):

$$S = \sum_{k=1}^{n} D1_k * D2_k * L1_k * L2_k, \qquad (6)$$

where n – a number of equivalent edges for current window and search request, $D1_k *$ $D2_k$ - relatedness factor for particular vertices pair, $L1_k$, $L2_k$ –weight s of k edges in $G_{request}$ and G_{sem} respectively.

The quality of the presented text comparison method was evaluated using large text collections satisfying the same search request on google.ru. Depending on the system settings, various values of the relevance factor were obtained. However, for collections that contain a query, or collections of similar content, the value of the coefficient was at least an order of magnitude higher than the value of the coefficient of other collections that are not similar in content. It was also shown that the deviation of the relevance factor increases linearly with increasing noise (Fig. 3), which proves that the comparison method works correctly and displays the correspondence between the collection window and the search query (Fig. 2).

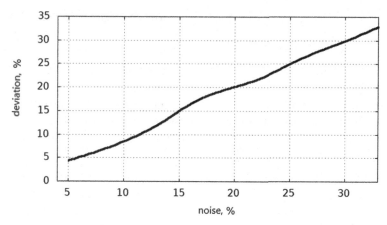

Fig. 2. The dependence between noise and relevance factor deviation.

3 Multi-layered Semantic Graph

Let us consider the analysis of professional dictionaries and the design of the respective semantic graphs. Explanatory dictionaries usually contain enough information to build the most important semantic dependencies. For example, the concept of *"bus"* is related to *"multi-seater"*, *"car"*, *"transportation"*, *"passenger"* and these vertices are connected to other concepts too.

The above examples of graph use show good results. Unfortunately, for scientific dictionaries, the analysis of the definition does not provide enough associative connections of the basic level. This is because each concept is associated with a large amount of specialized information that cannot be placed in the definition.

For example, the term *"integral calculus"* is defined as *"a section of mathematics that studies the concepts of integral, its properties and methods of computation"* [20]. It means that this concept should be connected with the concepts of *"mathematics"* and *"integral"* in the graph. At the same time, an attempt to introduce the word *"method"* into the graph, for which there is no own dictionary article, will lead to a huge number of links with various methods that are not related to integral calculus. In turn, the concept of *"integral"* according to the text of the definition has no connection with such fundamentally related concepts as *"primary function"* and *"indefinite integral"*, although in this case, through intermediate links, access to them can be obtained. At the same time, the full text of the relevant article in the mathematical encyclopedia contains information that leads to the direct connections.

The only way to overcome this problem is to create one or more layers $G_1, G_2, \ldots, G_n$ of contextual dependence over the graph G_0 based on more complete information from the encyclopedia. All further experiments were conducted with the Mathematical Encyclopedia [20], which contained about 7.5 thousand dictionary articles and a total of more than 2.6 million words. Table 3 provides statistics on the lengths of articles with examples. Almost half of the headings are represented by bigrams. Therefore, the information about the bigrams contained in the text should play a key role in the main topic analysis for encyclopedia articles.

Table 3. Statistics on article headings lengths presented in the Mathematical Encyclopedia.

Full text length	n = 1	n = 2	n = 3	n > 3
$2.6 \bullet 10^6$	12.0%	47.5%	17.4%	23.1%
Example	Первообразная (*primary function*)	Интегральное исчисление (*integral calculus*)	Линейное интегральное уравнение (*linear integral equation*)	Аналитическая теория дифференциальных уравнений (*analytical theory of differential equations*)

3.1 Layers Model

Obviously, the encyclopedia has much more information than intelligent dictionaries, but here we are faced with another problem – sometimes there is too much information related to the concept.

At the same time, encyclopedias are not only a source of highly professional information. They also have an undeniable advantage over disparate information collected from different sources. The main advantage in processing automation is that all entries in the encyclopedia are balanced both in volume and in the degree of content detail.

As noted, the basic layer of the semantic graph contains information extracted by general rules from the definitions of the concepts. The definition is usually contained in the first sentence of the corresponding entry. Next, each n-th layer is built for n-gram presented in the full text of the entry. Entries on average are quite extensive (on average 300–400 words), so it's reasonable to use only elements with maximum frequency for the layer, cutting off the least significant words on the threshold.

Thus, there is a problem of processing different articles with close statistical characteristics. Analysis of the entry headings has shown that it is possible to limit the processing of n-gram for n ≤ 3 (this approach allows to cover almost 77% of encyclopedia entries).

When working with n-grams for $n > 1$, you should take into account the specifics of the entry headings structure. In the Russian language, the order of words is quite free, as the morphological form of words allows us to determine their dependence. For example, the encyclopedia contains the entry named «*бесконечного порядка уравнение*» ("*infinite order equation*") but the concept «*уравнение бесконечного порядка*» ("*the equation of infinite order*") is much more often found in the full entry text. Such permutations are most typical for bigrams, compare: "*конечный автомат*" ("*state machine*") with "*автомат конечный*" ("*state machine*") and "*теория алгоритмов*" ("*theory of algorithm*"). with "*алгоритмов теория*" ("*theory of algorithm*").

Therefore, in this work, n-gram is not necessarily a strictly ordered sequence of n words. By n-gram we mean n words, which stand in the text side by side in some order.

In order to select n-grams carrying the main meaning within the entry, we need to find the set of most significant words. But we also should take into account the difference in the structure and contents of the texts. For this purpose, we will use cut-off thresholds for n-grams and bigrams based on the statistical characteristics of the articles. Particular threshold values can be calculated based on frequency data.

We propose to choose thresholds depending on the consistency factor of the lexical diversity of the text $d = l/m$, where l is full entry word count, m is unique canonical forms count. The more d, the less lexical diverse the text. Experiments conducted by the authors showed a good balance of processing of different texts using the following heuristic dependencies:

$$t_1/t_2 = f_1/f_2 * \sqrt{d_1/d_2}, \tag{7}$$

$$T(n)_1/T(n)_2 = \sqrt[n]{t_1/t_2}, \tag{8}$$

where t_i is a unigram threshold of i-th text, f_j is a maximum unigram frequency of the i-th text, d_i is a lexical diversity factor of the i-th text, $T(n)_i$ is a n-gram threshold of the i-th text, $n > 1$. Figures 3, 4, 5 demonstrate immediate neighbors (first-order neighbors) for the concept of "*integral calculus*" at $0 \leq n \leq 2$. It is obvious that as the layer number increases, the set of related concepts becomes more complete and more relevant.

Fig. 3. First-order neighbors for "*integral calculus*" concept in G_0.

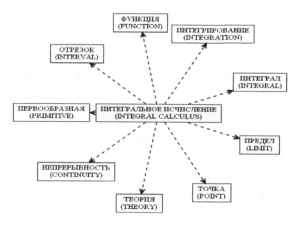

Fig. 4. First-order neighbors for "*integral calculus*" concept in G_1.

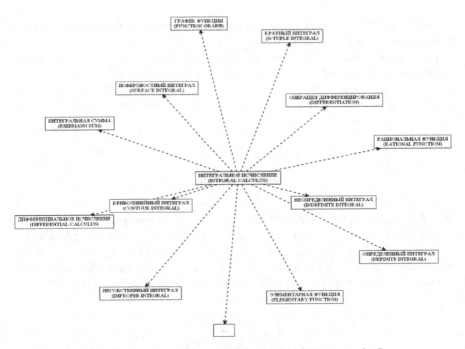

Fig. 5. First-order neighbors for *"integral calculus"* concept in G_2.

3.2 Text-To-Concept Algorithm

On each layer we calculate the statistical measure of the importance of the concept x for the document collection as a *TF-IDF* metric:

$$\beta_{TF-IDF}(x) = \beta_{TF}(x) * \beta_{IDF}(x), \tag{9}$$

where $\beta_{TF}(x)$ is a relative term frequency in the document, $\beta_{IDF}(x) = log(N/L(x))$ is inverse document frequency, N is number of documents in a collection, $L(x)$ – number of entries containing term x (equals to the number of inverse links from x vertex to G_1 vertices).

Therefore, the overall importance of term $\delta(x)$ can be obtained as follows:

$$\delta(x) = sum_{i=1}^{k}\beta_{TF-IDF}(x)^i, \tag{10}$$

where i is a layer index, k is a total number of layers.

The relevance factor for document v and some zero-layer concept can be calculated as a sum of $\delta(x)$ for each x term contained in document v.

Let's return to the multilayer for the generalized model. The domain independence of this model is due to the fact that it is used as a universal "model of the world."

Therefore, in the generalized model it is impossible to distinguish thematic concepts existing in a highly specialized domain, which we have considered on the example above.

In the world around us, because of its diversity, there is actually an infinite number of special domain areas. Therefore, it is impossible to build a basic semantic layer as a set of all known domain names. Processing data at one basic level allows us to solve only a limited number of tasks. For example, text categorizing based on statistical methods is difficult even with the use of limited semantics of the basic semantic graph. More information about the domain is needed.

We propose to use a huge amount of information accumulated in the Internet, for example, [21, 22], etc.

Domain-oriented texts are processed using the same principles of creating multi-layered graphs for $n \in 1.2, \ldots$, as described in Sect. 3.1. But the direction of layer processing is changed for the text relevance analysis phase. Due to the design features of the highly specialized graphs, the base layer corresponds to the semantic category.

For the general linguistic area, the base layer contains domain-independent information. In means that we can only obtain data on aspect dependencies after learning at the upper level of the graph layer. The authors' experience in the processing of the above sources has shown acceptable results already with the use of two layers on unigrams and bigrams. The additional third layer contains the highlighted aspects and is analogous to the base layer of the highly specialized mathematics domain.

Examples of the most significant aspects automatically highlighted for two domains are presented in Table 4.

Table 4. Examples of automatically highlighted aspect terms for "Movies" and "Restaurants" domains.

Domain	Movies	Restaurants
Aspect term examples	*Фильм (movie), время (time), герой (character), год (year), действие (action), жизнь (life), конец (end), момент (moment), работа (work), образ (character), сценарий (scenario)*	*Место (place), кухня (cousin), ресторан (restaurant), интерьер (interior), блюдо (dish/course), столик (table), обслуживание (service), впечатление (impression)*

4 Conclusion

The proposed method of using multi-layered graphs has been shown to be effective in both the analysis of domain-independent text documents and documents from a narrow domain area. The number of layers and their use are determined by the need to take into account the specifics of domain dependence. The method has a low time complexity and is aimed at a comprehensive solution to the problem of applying knowledge about language and the world to improve the quality of automatic text processing. Experiments confirm the applicability of the proposed formalized model.

References

1. Csurka, G., Dance, C., Fan, L., Willamowski, J., Bray, C.: Visual categorization with bags of keypoints. In: Proceedings of Workshop on Statistical Learning in Computer Vision (ECCV 2004), vol. 1, pp. 1–22 (2004)
2. Mikolov, T., Sutskever, I., Chen, K., Corrado, G., Dean, J.: Distributed representations of words and phrases and their compositionality. Adv. Neural. Inf. Process. Syst. **26**, 3111–3119 (2013)
3. Pennington, J., Socher, R., Manning, C.: Glove: global vectors for word representation. In: Proceedings of the 2014 Conference on Empirical Methods in Natural Language Processing (EMNLP), pp. 1532–1543. Association for Computational Linguistics, Doha, Qatar (2014)
4. Kutuzov, A., Kuzmenko, E.: Webvectors: a toolkit for building web interfaces for vector semantic models. In: AIST (2016)
5. Wang, J., Yang, Y., Mao, J., Huang, Z., Huang, C., Xu, W.: CNN-RNN: a unified framework for multi-label image classification. In: Proceedings of IEEE Conference on Computer Vision and Pattern Recognition (CVPR), pp. 2285–2294 (2016)
6. Akata, Z., Harchaoui, Z., Schmid, C.: Label-embedding for image classification. IEEE Trans. Pattern Anal. Mach. Intell. **38**, 1425–1438 (2015)
7. Alghunaim, A.: A vector space approach for aspect-based sentiment analysis. Ph.D. dissertation (2015)
8. Blinov, P., Kotelnikov, E.V.: Semantic similarity for aspect-based sentiment analysis. Russ. Digit. Libr. J. **18**, 120–137 (2015)
9. Ma, Y., Peng, H., Cambria, E.: Targeted aspect-based sentiment analysis via embedding commonsense knowledge into an attentive LSTM. In: AAAI (2018)
10. Baccianella, S., Esuli, A., Sebastiani, F.: Sentiwordnet 3.0: an enhanced lexical resource for sentiment analysis and opinion mining SentiWordNet. Analysis **10**, 1–12 (2010)
11. Cambria, E., Poria, S., Hazarika, D., Kwok, K.: Senticnet 5: discovering conceptual primitives for sentiment analysis by means of context embeddings. In: AAAI (2018)
12. Strapparava, C., Valitutti, A.: Wordnet-affect: an affective extension of wordnet. In: Proceedings of the 4th International Conference on Language Resources and Evaluation, Lisbon, pp. 1083–1086 (2004)
13. Berners-Lee, T., Hendler, J., Lassila, O.: The semantic web. a new form of web content that is meaningful to computers will unleash a revolution of new possibilities. Sci. Am. Mag. **284**, 1–5 (2001)
14. Krayvanova, V., Kryuchkova, E.: The mathematical model of the semantic analysis of phrases based on the trivial logic. In: Proceedings of Speech and computer (SPECOM), pp. 543–546 (2009)
15. Ozhegov, S., Shvedova, N.: Explanotary Dictionary of the Russian Language. http://lib.ru/DIC/OZHEGOW/
16. Abramov, N.: Dictionary of Russian Synonyms and words with close meanings. http://dict.buktopuha.net/data/abr1w.zip
17. Fellbaum, C. (ed.): WordNet: An Electronic Lexical Database. MIT Press, Cambridge (1998). ISBN 978-0-262-06197-1
18. Kazakov, M., Kryuchkova, E.: Classification of complex images based on semantic graph. J. Appl. Inform. **6**(54), 79–89 (2014)
19. Savchenko, V.: Semantic search algorithms in large text collections. In: Supplementary Proceedings of AIST, pp. 161–166 (2014)
20. Vinogradov, I.M. (ed.): Mathematical Encyclopedia in 5 volumes. Soviet Encyclopedia, Moscow (1977)

21. Pontiki, M., et al.: SemEval-2016 task 5: aspect based sentiment analysis. In: Proceedings of the 10th International Workshop on Semantic Evaluation (SemEval-2016), pp. 19–30. The Association for Computational Linguistics. San Diego, California (2016). https://doi.org/10.18653/v1/S16-1002
22. KinoPoisk. https://www.kinopoisk.ru/. Accessed 12 Nov 2020

Information and Computing Technologies in Automation and Control Science

Algorithmization of Autonomous Landing of Unmanned Aerial Vehicles by "Flexible" Kinematic Trajectories

Alexey Sergeev[1,2(✉)] ⓘ, Alexander Filimonov[3,4] ⓘ, and Nikolay Filimonov[1,2] ⓘ

[1] Trapeznikov Institute of Control Problems of RAS,
Profsoyuznaya street 65, 117997 Moscow, Russia
alxsrg95@gmail.com, nbfilimonov@mail.ru
[2] Lomonosov Moscow State University, Leninskie Gory 1, 119991 Moscow, Russia
[3] MIREA - Russian Technological University, Vernadsky prospect 78, 119454 Moscow, Russia
filimon_ab@mail.ru
[4] Moscow Aviation Institute (NRU), Volokolamskoe highw. 4, 125993 Moscow, Russia

Abstract. The final and one of the most important stages of the aircraft-type unmanned aerial vehicles (UAV) flight is landing. In this regard, the problem of automating the control by UAV landing in difficult meteorological conditions is becoming increasingly urgent. In some cases, for refueling and recharging UAVs it is advisable to use the dynamic mobile landing site (MLS) instead of the traditional stationary landing site (SLS). In the present paper, we consider the setting and solution of the control problem by the terminal landing maneuver of a UAV. It provides its transfer from the current initial state to the target final state along "flexible" kinematic trajectories both on the SLS and on the MLS. To solve the problem of the automatic landing of UAV on SLS or MLS the mathematical model of the dynamics of its movement was developed. It's based on the concept of "flexible" kinematic trajectories with spatial synchronization of controlled movements. The control algorithm by the terminal vertical landing maneuver of UAV on SLS by the method of dynamics inverse problem using the principle of "flexible" kinematic trajectories is developed. Also, the control algorithm by the terminal landing maneuver of UAV on MLS by the method of dynamics inverse problems using the principles of "flexible" kinematic trajectories and aiming into the target point was also developed. Computer approbation of the synthesized control algorithms for the landing maneuver of UAV «Aerosonde» under conditions of various wind disturbances was carried out using digital modeling in the Matlab environment.

Keywords: Landing maneuver · Stationary landing site · Mobile landing platform · Control algorithm · Method of dynamics inverse problems · Kinematic trajectories · Guidance to the target point

1 Introduction

In many spheres of man's vital activity, the unmanned aerial vehicles (UAV) are widely applied in both civil and military aviation [1]. They have some advantages as compared with the UAV of the helicopter type. They are the following: the greater radius

V. Jordan et al. (Eds.): HPCST 2020, CCIS 1304, pp. 181–200, 2020.
https://doi.org/10.1007/978-3-030-66895-2_12

of the action, longer time of the autonomous flight, the best aerodynamic indicators, greater coefficient of payload. They are very popular and effective as easy and inexpensive instruments for distant operations of search, rescuing, monitoring, observation inspection, scientific researches, patrolling, intelligence service, and so on. The landing is inalienable, final, and one of the most responsible, compound, and strain stage of the flight of arbitrary multiply vehicles. The most popular has the control problem by the trouble-free landing of UAV to the definite landing site in the automatic regime even in complex meteorological conditions [2]. The successful solution of the given problem has defined such important characteristics of UAV as flexibility, mobility, maneuverability, autonomy, and a low probability of damage and high multiplicity of the repeated application.

By now they developed the different methods for the solution of control problem by UAV on the static stationary landing site (SLS) chosen beforehand and equipped in the form of the land plot or the artificial landing runway (Fig. 1) being specially prepared. Here one can distinguish the control methods on the basis of GPS; the methods using the controllers with standard PID control laws; the control methods on the basis of dynamics inverse problems; the methods of linearization by feedback; the control method by sliding mode, the fuzzy and neural network control methods, the methods, using technical vision (see, for example [3–11]).

Fig. 1. Stationary landing site.

In some cases, is to be used the dynamical mobile landing site (MLS) in place of SLS for periodic add refueling and recharge UAV. The crane-beam [12], outstanding overboard of the moving craft-vehicle may be used such the landing platform for UAV of ship-based. The platform with stretched trapping net and device for capture, equipped on anyone motor transportation facilities (Fig. 2) may be used as the landing platform for the ground UAV [13].

The automatic control problem by UAV landing on MLS becomes more actual because its solution makes it possible to reduce the refueling time and to increase the coverage area of the refueling by fuel. As a result, it permits to increase the distance and period of the flight, to extend the sphere of the solved problems. Once the acceptable methods of solution of the given control problem by the landing are absent till now. In the present paper algorithms of automatic "soft" landing of UAV on SLS and MLS

Fig. 2. Dynamical mobile landing site

by the methods of dynamics inverse problems by "flexible" kinematic trajectories are suggested.

2   The Principle of "Flexible" Kinematic Trajectoriesin Control Problems by Terminal Maneuvers of UAV

In the theory and practice of the terminal control by the moving objects, the principle of "rigid" trajectories is dominated. This principle is implemented in the program-positional control strategy based on the conception of "unperturbed-perturbed movement" by Lyapunov – Letov. The given strategy is premised on the stabilization of the "rigid" nominal programmed trajectory of the object's movement planned beforehand. It provides the realization of terminal control objective in the "ideal conditions" (in the absence of the perturbation actions) of the movement. Recently, *the principle of the "flexible" trajectories,* based on an implementation of a pure programmed control strategy [14] becomes more widespread. The given principle is contained in the refusal from the binding of the controlled object's movement to the nominal trajectory planned beforehand. Here the mechanism of the formation of so-called "flexible" programmed trajectories of the movement is turned on the object's control loop. These trajectories are refreshed with the given periodicity and in this way are more useful. The given trajectories provide the realization of the terminal purpose by the object's control in "real conditions" of the movement (in the presence of perturbation actions) from the conditions, corresponding to the instant time of their renewal. This way the algorithm of programmed control is executed at each renewal interval that is the solution of the boundary-value problem of terminal control in "great" on the basis of the initial nonlinear model of the moving object's dynamics. At the same time, according to the terminology by N.N. Moiseev, the control way on the principle of "rigid" trajectories is named as the correction under the given program, and the control on the principle of "flexible" trajectories is named as the correction according to the final condition.

It should be noted that the conventional models of the UAV movement have a very general nature and do not take into account the main specific character of control problems by the moving objects, namely the kinematics of their movement into physical space.

The point is that the ordinary traditional way of the planning of terminal controlled trajectories UAV movement is based on the task of desired kinematic law of motion in the form of the dependence of space coordinates from the time:

$$\mathbf{r}(t), \quad t = \overline{t_0, \, t_f} \tag{1}$$

where $\mathbf{r} = (x_g, y_g, z_g)$ is vector of space coordinates; t_0 and t_f are instant time of the beginning and the end of the maneuver respectively. The given formalism leads to *nonstationary setting of control problem* and inclusion time factor t to the control algorithm being developed and so that the control process is found to be rigidly related to planned terminal instant time t_f. Because any program of purposeful movement UAV has the kinematic nature, then in the papers [14, 15] the new way of formalization of the trajectories of the device' movement was proposed and got the progression. The given way is based on their representation of the space curve in the surrounding physical space in the coordinate system $O_g X_g Y_g Z_g$. In this case, the desired trajectory of UAV may be given either by the system of functional equations:

$$\Phi_i(r) = 0, \quad i = \overline{1, 3} \tag{2}$$

or by parametrized function:

$$\mathbf{r}(\xi), \quad \xi = \overline{0, \xi_f} \tag{3}$$

where as the parameter ξ expediently to use one from the space coordinates of the device in the system of coordinates $O_g X_g Y_g Z_g$. It should be emphasized that in the task of the trajectory's UAV movement in the form of (1), so-called "time synchronization" of this trajectory is realized, and in the form of (2) and (3) its *"space synchronization"* is realized.

Thus, the realization of the principle of the "flexible" trajectories is possible as on the base of the mechanism of time synchronization of controlled movement so on the base of the mechanism of the space synchronization of controlled movements. The second way cardinally changes meaning tasks of control. The control processes are directed to the compulsion of the object to move along the given space curves, named further as *"kinematic trajectories"*.

3   Control by Vertical Landing Maneuver of UAV on the Stationary Landing Site

Let us consider UAV as the dynamic control object. For the description of its movement, we'll use the normal stationary system of coordinates $O_g X_g Y_g Z_g$ connected with the Earth, and the mobile system of coordinates $OXYZ$ connected with aerial vehicle (Fig. 3).

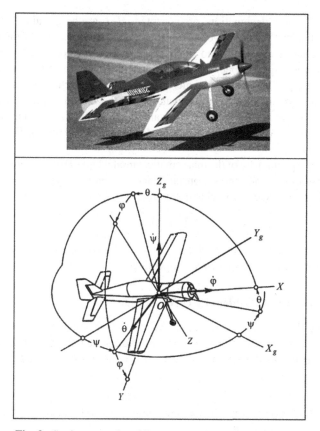

Fig. 3. Stationary and mobile systems of coordinates of UAV.

The mathematical model of the dynamics of vertical maneuver of UAV is described by the following system of differential equations [16]:

$$\dot{H} = V \, \sin \vartheta, \tag{4}$$

$$\dot{L} = V \, \cos \vartheta, \tag{5}$$

$$\dot{V} = (n_x(\alpha) - \sin \vartheta)g, \tag{6}$$

$$\dot{\vartheta} = \frac{(n_z(\alpha) - \cos \vartheta)g}{V}, \tag{7}$$

$$\dot{\theta} = q, \quad \theta = \alpha + \vartheta, \tag{8}$$

$$\dot{q} = \frac{q_D \bar{c} S}{I_y}(C_m(\alpha) + C_{m_q}\bar{c}q/V + C_{m_{\delta_e}}\delta_e). \tag{9}$$

Here the following designations are accepted: H, L are height and distance of the air part of the flight; V is the traveling speed; $\vartheta, \alpha, \theta$ are the angle of the trajectory's inclination, the angle of attack and pitch angle; n_x, n_z are longitudinal and normal overloads; q is the angular speed of pitch's change; δ_e is the angle of the elevator's deviation; $C_m, C_{m_q}, C_{m_{\delta_e}}$ are coefficients of aerodynamic pitch moment, longitudinal damping and the effectiveness of the elevator; I_y is the inertia moment relative to the transverse axis; $S, \bar{c}$ are the area of the wing and mean aerodynamic chord of the wing; q_D is velocity head: $q_D = \rho V^2/2$ (ρ is the air density); $g = 9.8$ m/s^2 is the acceleration of free incidence.

Let us note the peculiarity of the mathematical model (4)–(9) of the vertical maneuver of UAV. It's essentially a nonlinear, nonstationary, a dynamical system of the sixth order with state vector and the controlled action in the form of

$$\mathbf{x} = (L, H, \dot{L}, \dot{H}, \theta, q), \quad u = \delta_e,$$

functioning at the final time interval $[t_0, t_f]$, where t_0 and t_f are the moments of the beginning and the end of the maneuver respectively.

3.1 Setting of Control Problem by UAV Landing

The control problem by the landing maneuver of UAV consists in the synthesis of the deviation program of the elevator $\delta_e = \delta_e^*(t)$, providing the transfer of the aerial vehicle (4)–(9) from the arbitrary given initial state $\mathbf{x}(t_0) = \mathbf{x}_0$ in instant time $t = t_0$

$$L(t_0) = L_0, \; H(t_0) = H_0, \; \dot{L}(t_0) = \dot{L}_0, \; \dot{H}(t_0) = \dot{H}_0, \; \theta(t_0) = \theta_0, \; q(t_0) = q_0, \quad (10)$$

to terminal instant time $t = t_f$ for the given height H_f and distance L_f with the given meaning of vertical speed $\dot{H}_f$:

$$H(t_f) = H_f, \; L(t_f) = L_f, \; \dot{H}(t_f) = \dot{H}_f \qquad (11)$$

In the majority of traditional settings of terminal control problems, the terminal instant time t_f is fixed. But for the broad sphere of real terminal control problems by the moving objects, such restriction breaks down inexpedient or even inadmissible from point of view of the safety of controlled movements. The given restriction is inadmissible also for the formulated control problem by the landing maneuver of UAV. Its solution we'll construct with unfixed terminal instant time t_f on the basis of the principle of the "flexible" trajectories with the realization of space synchronization mechanism of controlled movements. In this connection we suppose that the desired trajectory of UAV landing is given in the form of the depending on the flight's height H from the distance $L \in [0, L_f]$:

$$H = H^*(\mathbf{x}_0, L), \qquad (12)$$

where $\mathbf{x}_0$ is the initial condition of the device for $L = 0$, where we suppose that the curve (12) is rather smooth. The controlling program of UAV we'll find as the function of the deviation of the elevator δ_e from the distance of the flight L:

$$\delta_e = \delta_e^*(\mathbf{x}_0, L), \qquad (13)$$

providing the movement of the device along the desired kinematic trajectory (12) and complying with the given boundary conditions (10).

The model of space synchronization of UAV movement in the vertical plane we'll get by means of exclusion time t from its initial model (4)–(9). We take the distance L in the capacity of the new independent variable. To this end let us divide every Eq. (4), (6)–(9) into the Eq. (5):

$$\frac{dH}{dL} = \tan \vartheta, \tag{14}$$

$$\frac{dV}{dL} = \frac{(n_x(\alpha) - \sin \vartheta)g}{V \cos \vartheta}, \tag{15}$$

$$\frac{d\vartheta}{dL} = \frac{(n_z(\alpha) - \cos \vartheta)g}{V^2 \cos \vartheta}, \tag{16}$$

$$\frac{d\theta}{dL} = \frac{q}{V \cos \vartheta}, \quad \theta = \alpha + \vartheta, \tag{17}$$

$$\frac{dq}{dL} = \frac{q_D \bar{c} S}{I_y V \cos \vartheta} (C_m(\alpha) + C_{m_q} \bar{c} q / V + C_{m_{\delta_e}} \delta_e). \tag{18}$$

The main difficulty of the solution of terminal control problem UAV, described by the model (14)–(18) is responsible for its two-point boundary nature, expressed by the conditions (10), (11). A very effective approach for its overcoming is based on the method of dynamics inverse problems [17]. It is premised on the task of the desired programmed object's movement, complying with the given boundary conditions at first, and then in the definition of the control, realizing this movement on the strength of the object's dynamics. Let us find the controlling action for UAV, providing the desired kinematic trajectories of its landing in a class of quasipolynomial functions in the form of the dependence of the flight's height H from the extension of the landing maneuver L:

$$H^*(L) = (H_0^* - H_f^*) \exp(-\lambda \tilde{L}) \times (1 + a_{H_1}\tilde{L} + a_{H_2}\tilde{L}^2)(1 - \tilde{L}) + H_f^*, \tag{19}$$

where

$$H^*(0) = H_0^*, \quad H^*(L_f) = H_f^*, \quad \lambda = 3/4, \quad \tilde{L} = \frac{L}{L_f},$$

$$a_{H_1} = \frac{dH^*}{dL}\bigg|_{L=0} = \tan \vartheta_0^*, \quad a_{H_2} = \frac{dH^*}{dL}\bigg|_{L=L_f} = \frac{\dot{H}_f^*}{V_f^*}.$$

3.2 The Development and Research of Control Algorithm by UAV Landing on SLS

For solving the considered control problem by vertical landing maneuver of UAV by "flexible" kinematic trajectories (19) algorithm [18] is worked out. It is based on the device of the dynamics inverse problems, represented in Table 1.

Table 1. Algorithm of control by UAV landing on SLS.

№ step	The sequence of computing action
1	To set the initial and final parameters of the landing maneuver UAV: $H_0, H_f, L_f, V_0, V_f, \dot{H}_f, \vartheta_0$
2	To calculate, according to (19), the kinematic trajectory of the landing $H^*(\mathbf{x}, L)$, complying with boundary conditions (10), (11) of the landing maneuver of UAV
3	To calculate, according to (14) the program of change of the inclination angle of the trajectory ϑ under the calculated program $H^*(\mathbf{x}, L)$: $\vartheta = \vartheta^*(\mathbf{x}, L)$
4	To calculate, according to (15) and (16) the programs of speed change V and the angle of attack α under the calculated programs $\vartheta^*(\mathbf{x}, L)$ and $\vartheta^*(\mathbf{x}, L)$: $V = V^*(\mathbf{x}, L), \alpha = \alpha^*(\mathbf{x}, L)$
5	To calculate, according to (17) the programs of change of pitch angle θ and angle speed q under the calculated programs $V^*(\mathbf{x}, L), \alpha^*(\mathbf{x}, L), \theta = \theta^*(\mathbf{x}, L)$: $\theta = \theta^*(\mathbf{x}, L)$, $q = q^*(\mathbf{x}, L)$
6	To calculate, according to (18) the unknown controlling program of change of elevator δ_e under the calculated programs $H^*(\mathbf{x}, L), \vartheta^*(\mathbf{x}, L), V^*(\mathbf{x}, L), \alpha^*(\mathbf{x}, L)$, $\theta = \theta^*(\mathbf{x}, L), q = q^*(\mathbf{x}, L)$: $\delta_e = \delta_e^*(\mathbf{x}, L)$

In the given algorithm, the realization of the strategy of the "flexible" kinematic trajectories is realized with the exchange from the initial condition $\mathbf{x}_0$ to the following $\mathbf{x}$:

$$H = H^*(\mathbf{x}, L) \Rightarrow \delta_e = \delta_e^*(\mathbf{x}, L),$$

as a result, the found algorithm of program-positional control realizes the mechanism of the feedback.

The synthesized algorithm is realized in MATLAB system and is approved by the example of the landing terminal maneuver of small-dimensional UAV "Aerosonde" [19, 20]. Its main parameters (weight, inertia, geometric, and aerodynamic) are presented in Tables 2, 3.

Table 2. Physical parameters of UAV "Aerosonde"

$m = 13{,}5\,\text{kg}$	$I_X = 0{,}824\,\text{kg*m}^2$
$S = 0{,}55\,\text{m}^2$	$I_Y = 1{,}135\,\text{kg*m}^2$
$\bar{c} = 0{,}19\,\text{m}$	$I_Z = 1{,}179\,\text{kg*m}^2$
$b = 2{,}9\,\text{m}$	$I_{XZ} = 1{,}120\,\text{kg*m}^2$

Table 3. Aerodynamic coefficients of UAV "Aerosonde".

$C_{L_0} = 0{,}23$	$C_{Y_0} = 0$	$C_{D_0} = 0{,}0434$
$C_{L_\alpha} = 5{,}616$ 1/rad	$C_{Y_\beta} = -0{,}83$ 1/rad	$C_{D_q} = 0$
$C_{L_q} = 7{,}95$ s/rad	$C_{Y_r} = 0$	$C_{D_{\delta_e}} = 0{,}13$ 1/rad
$C_{L_{\delta_e}} = 0{,}13$ 1/rad	$C_{Y_p} = 0$	$C_{n_0} = 0$
$C_{l_0} = 0$	$C_{Y_{\delta_a}} = -0{,}075$ 1/rad	$C_{n_\beta} = 0{,}0726$ 1/rad
$C_{l_\beta} = -0{,}13$ 1/rad	$C_{Y_{\delta_r}} = 0{,}1914$ 1/rad	$C_{n_p} = -0{,}069$ s/rad
$C_{l_p} = -0{,}5$ s/rad	$C_{m_0} = 0{,}135$	$C_{n_r} = -0{,}0946$ s/rad
$C_{l_r} = 0{,}25$	$C_{m_\alpha} = -2{,}73$ 1/rad	$C_{n_{\delta_a}} = 0{,}0108$ 1/rad
$C_{l_{\delta_a}} = -0{,}075$ 1/rad	$C_{m_q} = -38{,}2$ s/rad	$C_{n_{\delta_r}} = -0{,}693$ 1/rad
$C_{l_{\delta_r}} = 0{,}0024$ 1/rad	$C_{m_{\delta_e}} = -0{,}999$ 1/rad	

On the basis of the developed software, the analysis of the effectiveness of the synthesized control algorithm by the landing maneuver of UAV "Aerosonde" is conducted. In Figs. 4 and 5 the graphs of the "flexible" kinematic trajectories $H(L)$ and the deviations of the elevator $\delta_e(L)$ in the absence and in the presence of the constant wind perturbations for the following initial and terminal means of height, distance and speed of the flight are reduced:

$$L_0 = 0 \text{ m}, \ H_1 = 15 \text{ m}, \ H_2 = 18 \text{ m}, \ V_0 = 25 \text{ m/s},$$

$$L_f = 300 \text{ m}, \ H_f = 2 \text{ m}, \ V_f = 20 \text{ m/s}.$$

The approbation of the synthesized algorithm has shown the high quality of the landing maneuver of UAV with the realization of the requirements.

4   The Control of Spatial Landing Maneuver of UAV to the Moving Landing Site

The control problem by UAV landing on MLS is more compound than the problem by the landing on SLS discussed above. Let us consider the possibility of the dissemination of the described method for solving of control problem by the landing of UAV on SLS for the case of the autonomous landing on MLS, equipped on a motor vehicle (MV).

4.1   The Models of the Dynamics of UAV and MLS

We will consider UAV as the dynamic control object. The model of the dynamics of its space moving is described by the following system of the differential equations [21]:

$$\dot{x}_g = u \cos \theta \cos \psi + v(\sin \varphi \sin \theta \cos \psi - \cos \varphi \sin \psi)$$

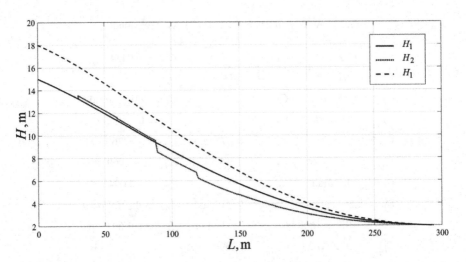

Fig. 4. Graphs of the "flexible" kinematic trajectories.

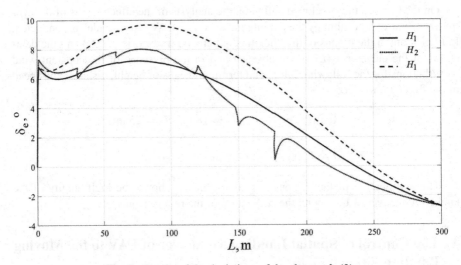

Fig. 5. Graphs of the deviations of the elevator $\delta_e(L)$.

$$+ w(\cos \varphi \sin \theta \cos \psi + \sin \varphi \sin \psi) + W_x, \tag{20}$$

$$\dot{y}_g = u \cos \theta \sin \psi + v(\sin \varphi \sin \theta \sin \psi + \cos \varphi \cos \psi)$$
$$+ w(\cos \varphi \sin \theta \sin \psi - \sin \varphi \sin \psi) + W_y, \tag{21}$$

$$\dot{z}_g = -u \sin \theta + v \sin \varphi \cos \theta + w \cos \varphi \cos \theta + W_z, \tag{22}$$

$$\dot{u} = rv - qw - g \sin \theta + \frac{q_D S C_D}{m} + \frac{P}{m}, \tag{23}$$

$$\dot{v} = pw - ru + g \cos\theta \sin\varphi + \frac{q_D S C_Y}{m}, \tag{24}$$

$$\dot{w} = qu - pv + g \cos\theta \cos\varphi + \frac{q_D S C_L}{m}, \tag{25}$$

$$\dot{p} = \frac{I_Y - I_Z}{I_X} qr + \frac{I_{XZ}}{I_X}(\dot{r} + pq) + \frac{q_D S b}{I_X} C_l, \tag{26}$$

$$\dot{q} = \frac{I_Z - I_X}{I_Y} pr + \frac{I_{XZ}}{I_Y}(r^2 - p^2) + \frac{q_D S \bar{c}}{I_Y} C_m, \tag{27}$$

$$\dot{r} = \frac{I_Z - I_X}{I_Z} pq + \frac{I_{XZ}}{I_Z}(\dot{p} - qr) + \frac{q_D S b}{I_Z} C_n, \tag{28}$$

$$\dot{\varphi} = p + q \sin\varphi \ \mathrm{tg}\theta + r \cos\varphi \ \mathrm{tg}\theta, \tag{29}$$

$$\dot{\theta} = q \cos\varphi - r \sin\varphi, \tag{30}$$

$$\dot{\psi} = q \sin\varphi \sec\theta + r \cos\varphi \sec\theta, \tag{31}$$

$$\alpha = \arctan\left(\frac{w}{u}\right), \quad \beta = \arcsin\left(\frac{v}{V}\right) \tag{32}$$

$$C_L = C_{L_0} + C_{L_\alpha}\alpha + C_{L_q}\bar{c}q/2/V + C_{L_{\delta e}}\delta e,$$

$$C_D = C_{D_0} + S/\pi/b^2 * (C_L)^2 + C_{D_q}\bar{c}q/2/V + C_{D_{\delta e}}\delta e,$$

$$C_m = C_{m_0} + C_{m_\alpha}\alpha + C_{m_q}\bar{c}q/2/V + C_{m_{\delta e}}\delta e,$$

$$C_Y = C_{Y_0} + C_{Y_\beta}\beta + C_{Y_p}bp/2/V + C_{Y_r}br/2/V + C_{Y_{\delta a}}\delta a + C_{Y_{\delta r}}\delta r,$$

$$C_l = C_{l_0} + C_{l_\beta}\beta + C_{l_p}bp/2/V + C_{l_r}br/2/V + C_{l_{\delta a}}\delta a + C_{l_{\delta r}}\delta r,$$

$$C_n = C_{n_0} + C_{n_\beta}\beta + C_{n_p}bp/2/V + C_{n_r}br/2/V + C_{n_{\delta a}}\delta a + C_{n_{\delta r}}\delta r.$$

where

- x_g, y_g, z_g are coordinates of the device relative to the Earth coordinates system, $O_g X_g Y_g Z_g$;
- $V_g = \sqrt{\dot{x}_g^2 + \dot{y}_g^2 + \dot{z}_g^2}$ is ground speed; u, v, w are speeds in the connected coordinates systems $OXYZ$;
- W_x, W_y, W_z are the views of the wind speed in the coordinates system $O_g X_g Y_g Z_g$;
- θ, φ, ψ and p, q, r are the pitch angles, angles of bank, angles of yaw and their angle speed respectively;

- P is thrust force of the motors; $V = \sqrt{u^2 + v^2 + w^2}$ is the airspeed;
- I_X, I_Y, I_Z, I_{XZ} are the moments of inertia relative to the principal axes;
- $S, \bar{c}, b$ are the area of wing, mean aerodynamic chord of the wing, and the range of the wing respectively;
- α and β are the attack angle, and sliding angle respectively;
- C_D, C_Y, C_L are aerodynamic coefficients of forces;
- C_l, C_m, C_n are aerodynamic coefficients of the moments;
- q_D is pressure head ($q_D = 0.5\ \rho V^2$, $\rho = 1.225$ kg/m^3 is the air density); $g = 9.8$ m/s^2 is the acceleration of free incident.

At state vector of UAV, we will consider the vector in the form of

$$\mathbf{x} = (x_g, y_g, z_g, \dot{x}_g, \dot{y}_g, \dot{z}_g, \theta, \varphi, \psi, p, q, r),$$

and in the capacity of its controlling variables let us examine vector:

$$\mathbf{u} = (\delta_e, \delta_r, \delta_a, P),$$

where $\delta_e, \delta_r, \delta_a$ are the angles of the deviation of elevators, the direction and bank (ailerons) respectively.

Let us consider MLS in the form of a horizontal landing catching net, equipped on the roof of a microbus. In doing so for the safety of the landing maneuver of the device we'll choose a rectilinear horizontal section of the locality all along x_{max}, irregularities of which may be compensated by the hangers of the platform and MV. Let us suppose, that VM doesn't maneuver moving rectilinearly and uniformly. In doing so, the landing trajectory of the UAV has a small side deviation from the vertical plane where the trajectory of moving MV ($y_g << z_g << x_{max}$) is positioned. The standard stationary coordinates system $O_g X_g Y_g Z_g$, connected with the Earth, and movement a mobile coordinates system $OXYZ$ connected with MV are used for the description of MLS (Fig. 6).

The model of the dynamics' movement of MLS is described by the following system of the differential equations:

$$\dot{x}_{pl} = V_{pl}, \tag{33}$$

$$\dot{y}_{pl} = 0, \quad \dot{z}_{pl} = 0 \tag{34}$$

where x_{pl}, y_{pl}, z_{pl} and V_{pl} are the coordinates and speed of the platform's movement respectively: they define its state vector: $\mathbf{x}_{pl} = (x_{pl}, y_{pl}, z_{pl}, V_{pl})$.

4.2 Setting of the Control Problem by UAV Landing on MLS

The control problem by the landing maneuver of UAV on MLS consists in the synthesis of the deviations programs of elevators, direction, bank, and thrust changes:

$$\delta_e = \delta_e^*(t), \quad \delta_a = \delta_a^*(t), \quad \delta_r = \delta_r^*(t), \quad P = P^*(t),$$

providing the transfer of the aerial vehicle (20)–(32) from the arbitrarily given initial state $\mathbf{x}_0$ at the instant time $t = t_0$ to the final state $\mathbf{x}_f$ at the terminal instant time $t = t_f$ with the admissible terminal error in the moment of the device's approach with MLS (33), (34):

$$|x_{g_f} - x_{\mathrm{pl}_f}| \leq \varepsilon_x^*, \tag{35a}$$

$$|y_{g_f} - y_{\mathrm{pl}_f}| \leq \varepsilon_y^*, \quad |z_{g_f} - z_{\mathrm{pl}_f}| \leq \varepsilon_z^*, \quad |V_{g_f} - V_{\mathrm{pl}_f}| \leq \varepsilon_V^*, \tag{35b}$$

where $x_{\mathrm{pl}_f}, y_{\mathrm{pl}_f}, z_{\mathrm{pl}_f}, V_{\mathrm{pl}_f}$ are the final means of the variables of the platform's state; ε_x^*, $\varepsilon_y^*, \varepsilon_z^*, \varepsilon_V^*$ are the limited admissible means of vector components of a terminal error of the UAV $\boldsymbol{\varepsilon} = (\varepsilon_x, \varepsilon_y, \varepsilon_z, \varepsilon_V)$.

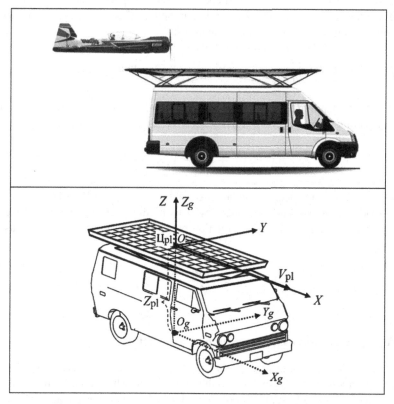

Fig. 6. Stationary and mobile systems of coordinates of MLS

Let us reduce the solution of the formulated control problem by the landing of UAV by the method of dynamics inverse problems with the use of the principles of "flexible" kinematic trajectories and guidance to the target point.

4.3 The Principle of "Flexible" Kinematic Trajectories

Following the given principle, the kinematic trajectories of the landing maneuver of UAV let us compose a class of quasipolynomial functions:

$$z^*(x) = (z_0 - z_f) \exp(-\lambda_z x')(1 + a_{z_1} x' + a_{z_2} x'^2)(1 - x') + z_f, \qquad (36)$$

$$y^*(x) = (y_0 - y_f) \exp(-\lambda_y x')(1 + a_{y_1} x' + a_{y_2} x'^2)(1 - x') + y_f, \qquad (37)$$

$$x' = \frac{x - x_0}{x_f - x_0},$$

where the variables x_0, y_0, z_0 and x_f, y_f, z_f are coordinates of the initial and finish of the landing maneuver of aerial vehicle respectively. Here the parameters λ_y, λ_z are given, and the coefficients $a_{z_1}, a_{z_2}, a_{y_1}, a_{y_2}$ are determined by the initial and finish states of UAV from the following system of the equations:

$$\left.\frac{dz^*}{dx}\right|_{x=0} = \tan\vartheta_0, \quad \left.\frac{dz^*}{dx}\right|_{x=x_f} = \tan\vartheta_f, \quad \left.\frac{dy^*}{dx}\right|_{x=0} = \tan\xi_0, \quad \left.\frac{dy^*}{dx}\right|_{x=x_f} = \tan\xi_f,$$

For the "soft" landing of UAV on MLS its travelling speed V_g is bound to be lowered till the platform's speed V_{pl} ($V_g \geq V_{pl}$):

$$V_g^*(x) = V_{g0} + (V_{pl} - V_{g0})\frac{x - x_0}{x_f - x_0}, \qquad (38)$$

The trajectories and speed of the landing maneuver of UAV (36)–(38) are the functions of space variable x, and the equations of its moment (20)–(32) are the functions of a temporary variable t. For the correlation between these variables we may use the following relations:

$$\frac{dy}{dx} = \frac{dy/dt}{dx/dt} = \frac{\dot{y}_g}{\dot{x}_g}, \quad \frac{dz}{dx} = \frac{dz/dt}{dx/dt} = \frac{\dot{z}_g}{\dot{x}_g}, \quad V_g^2 = \dot{x}_g^2 + \dot{y}_g^2 + \dot{z}_g^2. \qquad (39)$$

4.4 The Principle of the Guidance to the Target Point

For calculation of the coordinate x_f that is the target point of contact of UAV with MLS to the principle of the guidance to the target point, which realizes the method of "pursuit with look-ahead" [22] is used. According to the given principle the trajectory of the landing maneuver of the aerial vehicle is conducted. It provides its recasting in terminal instant time $t = t_f$ to the given neighborhood $\varepsilon^* = (\varepsilon_x^*, \varepsilon_y^*, \varepsilon_z^*, \varepsilon_V^*)$ of the target point of the meeting with a platform which is determined by the admissible terminal error (35). The decrease of value of the given error is achieved by the calculation of the planning target point of the meeting UAV and MLS in every instant time of the renewal of a "flexible" landing kinematic trajectory. For this case on the basis of the calculated programmed landing trajectories $z^*(x)$, $y^*(x)$, program $V_g^*(x)$ and the relations (39) the planning terminal time of the landing maneuver t_f and coordinate of MLS x_{pl_f} are calculated:

$$x_{pl_f} = x_{pl_0} + V_{pl}t_f. \tag{40}$$

Then based on the comparison of terminal means of coordinates UAV x_f and MLS x_{pl_f} the target end state of the device $\mathbf{x}_f$ is determined:

- if the condition is good (35a) so the calculated final state vector $\mathbf{x}_f$, landing trajectories $z^*(x)$, $\mathbf{x}$ and the program $V_g^*(x)$ of UAV are chosen.
- if the condition is not good (35a) so the final coordinate of UAV x_f is assigned a value of the final coordinate of the platform x_{pl_f} (40) and recalculation of the target final state vector $\mathbf{x}_f$ is performed.

Thus, in the basis of the principle of the guidance to the target point $\mathbf{x}_f$ of the meeting UAV with MLS, the realization of the condition of terminal error (35a) is situated.

4.5 Control Algorithm by the Landing of UAV

For the solution of the considered control problem by UAV landing along "flexible" kinematic trajectories (36), (37) the algorithm [23] is developed. It's based on the method of the dynamics inverse problems presented in Table 4.

Table 4. Control algorithm by the landing of UAV.

№ step	Sequence of calculations
1	*The setting of the initial and terminal states*
1.1	To set the initial states of UAV $\mathbf{x}_0$ and MLS $\mathbf{x}_{pl_0}$
1.2	To set the final coordinates x_f, y_f, z_f and speed V_{g_f} of UAV: $x_f = (x_{max} - x_{pl_0})/2, y_f = y_{pl_f}, z_f = z_{pl_f}, V_{g_f} = V_{pl}$
1.3	To set time mesh width Δt
2	*The search of coordinate target point of the meeting x_f and the calculation of the programs landing trajectories $z^*(x), y^*(x)$*
2.1	To assume the number of the iteration $k = 1$
2.2	To assume $i = 0$
2.3	To calculate x_{i+1} on $x_i, \dot{x}_{g_i}: x_{i+1} = x_i + \dot{x}_{g_i}\Delta t$
2.4	If $x_{i+1} \leq x_f$, pass to step 2.4.1, otherwise pass to step 2.4.2
2.4.1	To calculate y_{i+1}, z_{i+1} supposing $x = x_{i+1}$ in (36), (37) and pass to step 2.9
2.4.2	To calculate $t_f: t_f = \Delta t * i$ and pass to step 2.5
2.5	To calculate x_{pl_f} on t_f according to (40)

(continued)

Table 4. (*continued*)

№ step	Sequence of calculations
2.6	If the inequality (35a) is fulfilled for $x_{g_f} = x_i$ pass to step 2.7, otherwise pass to step 2.6.1
2.6.1	To assume $x_f = x_{pl_f}$, $k = k + 1$ and pass to step 2.2
2.7	If the inequality (35b) is fulfilled for $y_{g_f} = y_i$, $z_{g_f} = z_i$, $V_{g_f} = V_{gi}$, pass to step 3, otherwise pass to step 2.8
2.7.1	If $k \leq k_{max}$ pass to step 2.6.1, otherwise stop the calculation and interrupt the landing maneuver of UAV from behind its non-fulfilment for the given initial and final conditions
2.8	To calculate $V_{g_{i+1}}$ for $x = x_{i+1}$ according to (38)
2.9	To calculate $\dot{x}_{g_{i+1}}, \dot{y}_{g_{i+1}}, \dot{z}_{g_{i+1}}$ for $V_g = V_{g_{i+1}}$, $dx = x_{i+1} - x_i$, $dy = y_{i+1} - y_i$, $dz = z_{i+1} - z_i$, $dt = \Delta t$ according to (39)
2.10	To set $\theta_i, \varphi_i, \psi_i$: $\theta_i \in [2°, 4°]$, $\varphi_i = 0$, $\psi_i = \arctan((y_{i+1} - y_i) / (x_{i+1} - x_i))$
2.11	To assume $i = i + 1$ and pass to step 2.3
3	***The calculation of the unknown controlling programs*** $\delta_e^*, \delta_a^*, \delta_r^*, P^*$ (all calculations, connected with the differential Eqs. (20)–(32) are constructed by the corresponding finite-difference analogs by Euler)
3.1	To assume $j = 0$
3.2	If $x_j < x_f$ pass to step 3.3, otherwise stop calculations in connection with found unknown controlling programs
3.3	According to the difference analogs of the Eqs. (20)–(22) calculate $u_{j+1}, v_{j+1}, w_{j+1}$ on $x_{j+2}, x_{j+1}, y_{j+2}, y_{j+1}, z_{j+2}, z_{j+1}$
3.4	According to difference analogs of the Eqs. (29)–(31) calculate $p_{j+1}, q_{j+1}, r_{j+1}$ on $\theta_{j+2}, \theta_{j+1}, \varphi_{j+2}, \varphi_{j+1}, \psi_{j+2}, \psi_{j+1}$
3.5	According to the difference analogs of the Eqs. (23)–(28) calculate $\delta_{e_j}^*, \delta_{a_j}^*, \delta_{r_j}^*, P_j^*$ on $u_{j+1}, u_j, v_{j+1}, v_j, w_{j+1}, w_j, p_{j+1}, p_j, q_{j+1}, q_j, r_{j+1}, r_j$
3.6	To assume $j = j + 1$ and pass to step 3.2

It should be pointed out that the program controlling actions $\delta_e^*, \delta_a^*, \delta_r^*, P^*$ are synthesized on the basis of the reduced algorithm. They realize the open control by the landing of UAV on MLS. But executing the change of the initial states $\mathbf{x}_0$ and MLS $\mathbf{x}_{pl_0}$ to the flowing $\mathbf{x}, \mathbf{x}_{pl_0}$ and taking into account, that the modern fast board digital calculating computer permit to form "flexible" landing trajectories on every step of quantification time Δt. The given programmed strategies of control, in fact, realize the feedback mechanism.

4.6 Computer Approbation of the UAV Landing Control Algorithm

For the approbation of the developed algorithm, the computing simulation of the landing maneuver on UAV of the airplane type "Aerosonde" on MLS was considered. The simulation was fulfilled in MATLAB medium with the use of the developed software. The analysis of the algorithm's effectiveness is conducted for the landing maneuver of UAV from height $z_0 = 15$ m on MLS located at the height $z_{pl} = 2$ m above the Earth and moving with speed $V_{pl} = 20$ m/s. The different initial speeds of UAV V_{g_0} ($V_1 = 25$ m/s and $V_2 = 28$ m/s) and the initial positions of the platform x_{pl_0} ($r_1 = 30$ m and $r_2 = 50$ m) in the absence of (W_1) and in the presence of the harmonic (W_2) and constant (W_3) wind perturbations were considered. At the renewal moment for the landing kinematic trajectory of UAV the random perturbation of its condition, and also the conditions of MLS were given. In Figs. 7, 8, 9 and 10 the results of the computer simulation of the land are represented.

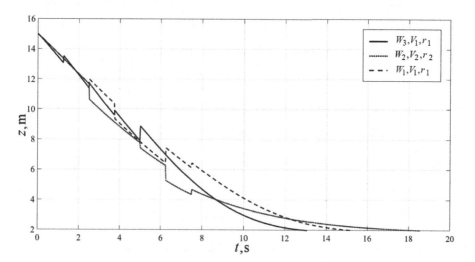

Fig. 7. Graphs of the landing trajectories $z(t)$.

The graphs of the landing trajectories $z(t)$ and $z(x)$ are in Fig. 7 and Fig. 8 respectively. The graphs of alterations of the target point $x_f(x)$ are in Fig. 9. The graphs of the controls of the elevator $\delta_e(x)$ are in Fig. 10.

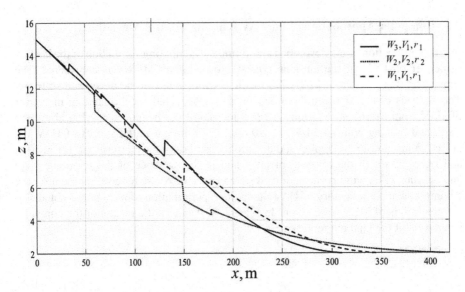

Fig. 8. Graphs of the landing trajectories $z(x)$.

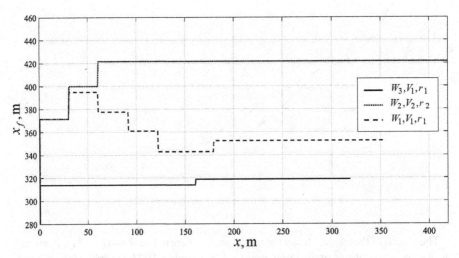

Fig. 9. Graphs of the alterations of the target point $x_f(x)$.

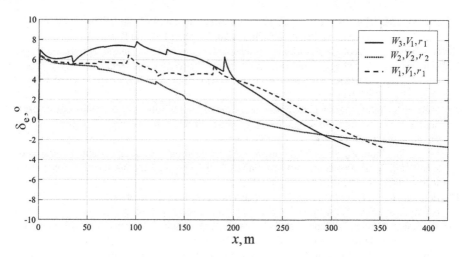

Fig. 10. Graphs of the controlling actions of elevator $\delta_e(x)$

5 Conclusion

The actual problems of the synthesis of the control algorithm by the landing terminal manoeuver of UAV on the stationary landing site and the moving landing platform are set and solved. The corresponding control algorithms by UAV "Aerosonde" are developed and approved in MATLAB medium. The getting results of the computer simulation have shown the great effectiveness of the synthesized control algorithms by means of all requirements and restrictions.

References

1. Unmanned Aerial Vehicles: Reference Guide. Izdatelstvo Poligraficheskij centr «Nauchnaya kniga», Voronezh (2015). (in Russian)
2. Beard, R.W., McLain, T.W.: Small Unmanned Aircraft: Theory and Practice. Princeton University Press, Princeton (2012)
3. Gautam, A., Sujit, P.B., Saripalli, S.: A survey of autonomous landing techniques for UAVs. In: Proceedings of the 2014 International Conference on Unmanned Aircraft Systems (ICUAS), Orlando, FL, USA, pp. 1210–1218 (2014)
4. Gommer, A.S.: Review of methods of autonomous landing of UAVs. In: Science-Intensive Research as the Basis of Innovative Development of Society: Collection of Articles of the International Scientific and Practical Conference, vol. 2, pp. 70–76. OMEGA SCIENCE, UFA (2019). (in Russian)
5. Ageev, A.M., Belyaev, V.V., Bondarev, V.G., Protsenko, V.V.: The Systems of Automatic Landing for Unmanned Aerial Vehicles. Probl. Soln. Mil. Thought **4**, 130–136 (2020). (in Russian)
6. Kravchenko, P.P., Kulikov, L.I., Scherbinin, V.V.: Application of the method of optimized delta-transformations in the control landing problem for an unmanned aerial vehicle. J. Comput. Syst. Sci. Int. **58**(5), 786–800 (2019). (in Russian)

7. Penyaz, I.M.: The innovative developments concerning a way of soft landing of the modern unmanned aerial vehicles. Probl. Saf. Flights **10**, 34–38 (2019). (in Russian)
8. Burchett, B.T.: Feedback linearization guidance for approach and landing of reusable launch vehicles. In: Proceedings of the American Control Conference, pp. 2093–2097. IEEE (2005)
9. Nho, K., Agarwal, R.K.: Automatic landing system design using fuzzy logic. J. Guid. Control Dyn. **23**(2), 298–304 (2000)
10. Malaek, S., Sadati, N., Izadi, H., Pakmehr, M.: Intelligent autolanding controller design using neural networks and fuzzy logic. In: Proceedings of 5th Asian Control Conference, vol. 1, pp. 365–373. IEEE (2004)
11. Wang, R., Zhou, Z., Shen, Y.: Flying-wing UAV landing control and simulation based on mixed H_2/H_∞. In: Proceedings of International Conference on Mechatronics and Automation, ICMA 2007, pp. 1523–1528. IEEE (2007)
12. Podoplekin, Yu.F., Sharov, S.N.: Key aspects of theory and design of landing systems of UAV on small vessels. Informacionno–Upravlyayushchie sistemy **6**, 14–24 (2013)
13. Hérissé, B., Hamel, T., Mahony, R., Russotto, F.-X.: Landing a VTOL unmanned aerial vehicle on a moving platform using optical flow. IEEE Trans. Rob. **28**(1), 77–89 (2012)
14. Filimonov, A.B., Filimonov, N.B.: Methods of «Flexible» Trajectories in the Tasks of Terminal Control of Aircraft Vertical Maneuvers; the Monograph «Problems of Control of Complex Dynamic Objects of Aviation and Space Technology», vol. 2. Machinostroenie, Moscow (2015). (in Russian)
15. Teryaev, E.D., Filimonov, A.B., Filimonov, N.B., Petrin, K.V.: The conception of «Flexible Kinematic Trajectories» in the problems of terminal control by moving objects. Mekhatronika, Avtomatizaciya, Upravlenie **12**, 7–15 (2011). (in Russian)
16. Byushgens, G.S.: Dynamics of the Airplane. Dynamics of Longitudinal and Transverse Motion. Machinostroenie, Moscow (1979). (in Russian)
17. Krutko, P.D.: Inverse Problems of Dynamics in the Theory of Automatic Control. Mashinostroenie, Moscow (2004). (in Russian)
18. Filimonov, N.B., Sergeev, A.A.: Synthesis of control algorithm for the UAV vertical landing maneuver by the method of flexible kinematic trajectories. J. Adv. Res. Tech. Sci. **17**(2), 150–156 (2019). (in Russian)
19. Burston, M.T., Sabatini, R., Clothier, R., Gardi, A.: Reverse engineering of a fixed wing unmanned aircraft 6-DoF model for navigation and applications. Appl. Mech. Mater. **629**, 164–169 (2014)
20. Bateman, F., Noura, H., Ouladsine, M.: Fault diagnosis and fault-tolerant control strategy for the aerosonde UAV. IEEE Trans. Aerosp. Electron. Syst. **47**(3), 2119–2137 (2011)
21. Byushgens, G.S., Studnev, R.V.: Dynamics of the Airplane. Spatial Motion. Machinostroenie, Moscow (1983). (in Russian)
22. Panjkov, S.J., Zaburaev, J.E., Matveev, A.M.: Theory and Methods of Aircraft Control, vol. 1. UVAU GA, Ulyanovsk (2006). (in Russian)
23. Sergeev, A.A., Filimonov, N.B.: Controlling the maneuver of an unmanned aerial vehicle when landing on a mobile platform using the method of "Flexible" kinematic trajectories. J. Instrum. Eng. **63**(9), 803–812 (2020). (in Russian)

Features of Designing a Variable-Frequency Electric Drive Control System with a Microprocessor-Based Sinusoidal Signal Generator

Ishembek Kadyrov[1]([⊠]) [iD], Nurzat Karaeva[1], Zheenbek Andarbekov[1], and Kyyal Kadyrkulova[2]

[1] Skryabin Kyrgyz National Agrarian University, Mederov Street 68, 720005 Bishkek, Kyrgyz Republic
bgtu_kg@mail.ru
[2] Razzakov Kyrgyz State Technical University, Ch. Aitmatov Ave. 66, 720044 Bishkek, Kyrgyz Republic

Abstract. The work deals with the accuracy requirements for the manufacture of large-size body parts. The article substantiates the possibility of using the variable-frequency electric drive of the hydraulic directional valve in the servo system of the movement and control of the working member in ultra-high-pressure hydraulic presses. The authors consider the problem of finding the frequency converters that supply induction motors. It has been noted that the most acceptable one is the cycloconverter (CCV), consisting of three reversible DC bridge converters. The authors also have found that if the quality of the refinement of controlling effects on the electric drive control system is satisfactory, a significant influence on the accuracy of the formation of induction motor phase currents is produced by the perturbing effect from the EMF of IM rotation, whose schematic diagrams are presented. It has been proposed to analyze the requirements to the generator dictated by the IM frequency regulation law in vector control, when the rotor flux linkage must be maintained at a constant level. In conclusion it has been pointed out that it is necessary to create a three-phase sinusoidal symmetrical voltage generator with a complex algorithm that implements a certain system of equations for harmonically varying voltages. The functional schematic diagram of the electric drive using the CCV-IM system has been presented. The microprocessor has been selected, its hardware framework has been described; a block diagram has been created, used to execute the software part of the MPSSG work.

Keywords: Hydraulic directional valve · Cycloconverter · Induction motor · Tachogenerator · Thyristor converter · Microprocessor · Calculation algorithm · Block diagram

1 Introduction

The main task of each product manufacturing technological process is to ensure its high quality at the lowest cost. Modern products of machine-building industry, based

V. Jordan et al. (Eds.): HPCST 2020, CCIS 1304, pp. 201–217, 2020.
https://doi.org/10.1007/978-3-030-66895-2_13

on their intended service, have increasing requirements to their quality and especially to one of its components - the accuracy. This is especially relevant for machine-building production, where the technological process includes such operations as forging, bulk forming, extruding, draw-forming, piercing, cutting and other processes. In order to perform production operations, e.g. cold extrusion, ultra-high-pressure hydraulic presses are used, whose working member is driven by a hydraulic drive, whose advantages are well known.

Figure 1 shows a hydraulic schematic diagram of a three-cylinder press, which uses mineral oil as the working fluid in the hydraulic cylinder, which is pumped using high-speed pumps [1, 2].

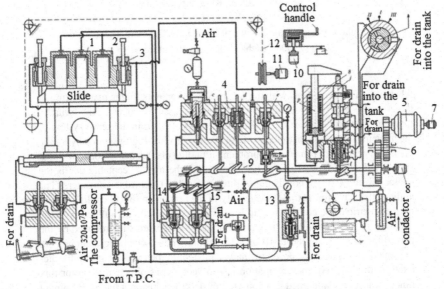

Fig. 1. Hydraulic schematic diagram of control of the three-cylinder forging press with two stages of reinforcement: 1 - central working cylinder; 2 - side working cylinders; 3 - pullback cylinders; 4 - main directional valve; 5 - induction motor; 6 - reduction gear; 7 - tachogenerator; 8 - synchro receiver; 9 - shaft of the hydraulic directional valve; 10 - synchro transmitter; 11 - synchro receiver; 12 - system that converts the forward movement of the slide into the rotational movement of the synchro receiver; 13 - filler; 14 - central cylinder filler-drain valve; 15 - side cylinders filler-drain valve.

A characteristic feature in the construction of a hydraulic press control system is the large dimensions of the processed products. Therefore, in these units, control flexibility can only be achieved by using reliable hydraulic equipment, a sensitive servo system, with the help of which remote and active automatic control of the size of the manufactured product is built [3–5].

The paper [6] describes the construction procedure of the servo system for the control of the ultra-high-pressure hydraulic presses; the elements enabling effective control of the object are given. Schematic solutions showing the principle of working out the commands set by manipulation actions of control members are shown, so that the working member of a hydraulic press would provide the work by cycles formulated during the description of a hydraulic press [7]. These schematic solutions are as follows:

1. holding the slide overhanging during the workpiece and tool installation;
2. providing the fast downward movement mode of the slide until the tool comes into contact with the workpiece;
3. specifying the working stroke, ensuring the main stroke of the technological process, which results in achieving a set shape of the product;
4. fixing the working member for a certain time to obtain a good quality of the surface of the finished part, i.e. of a given purity, uniformity of structure, etc.;
5. providing the backward stroke to the starting point in fast movement mode.

This paper also proposes the features of building a hydraulic directional valve electric drive control system, shown in Fig. 1. Its main elements are an induction motor (IM) with a short-circuit rotor 5 that triggers the shaft of the hydraulic directional valve 9; a cycloconverter consisting of reversible thyristor DC converters that ensure the provision of power to the IM stator windings. In addition, there are the controlled parameters sensors of the electric drive; the current and speed controllers with their adjustment to a modular optimum, which form the mechanical characteristics of the electric drive for accurate refining of the working member in cycles.

When building variable-frequency electric drive control systems we focused on the use of modern microprocessor-based hardware components of control, given that the static and dynamic properties of the entire system depend on the quality and accuracy of controlling effects formation. Moreover, the microprocessor version of the sinusoidal signal generator (SSG) has already been considered in the paper [8].

However, the materials of the paper [8] are outdated. For example, developers do not use the principle of construction of digital-to-analog converters given in the paper [8] at this stage; besides, the programming algorithm has also changed.

Before proceeding to the development of the next version of the MPSSG, which includes a new principle of pulse-width modulation using modern microcontrollers, it is necessary to describe the features of hydraulic press control in more detail. The need for this is explained by the following factors. First, after studying the main operating modes performed by the working member, we can list the requirements for the electric drive of the shaft of the hydraulic directional valve. Second, the features of the induction motor phase currents formation will be taken into account, primarily for operations of such draw forming of body parts.

In accordance with the description given in [7], the setting for the step cycle of the working member movement is performed using the control handle (Fig. 1). The rotation of the handle leads to the formation of a mismatch between two devices. The first is synchro transmitter (ST) 10 mounted on the shaft of the handle, and synchro receiver 11. The shaft is connected with the slide through a system that converts the forward movement of the slide into rotational movement of synchro receiver 12 and controls the positioning of the working member. The second is ST 10 and synchro receiver 13 mounted on the shaft of the hydraulic directional valve and controlling the speed of refining this cycle.

The hydraulic press slide moves forward when the hydraulic directional valve shaft rotates due to the opening of valves and throttles installed on main directional valve 4, when the liquid from the lifting cylinders is forced into the central cylinder of press 1 and the slide is lowered. Moreover, central cylinder 1 and side cylinders 2 are filled with a liquid from filler 13 through filler-drain valves 14–15, which open due to the pressure drop.

The electric drive using the CCV-IM system, as a power element, is included in the servo system to ensure the required speed of movement of the working member. At the same time, the range of IM speed control depends on the speed of the slide movement, from high-speed when the slide is in free movement to a very slow movement when refining the technological process, for example, draw forming of a body part. Thus, the main requirements for the electric drive are:

1. provision of a wide range of induction motor speed control;
2. provision of a high value of rigidity of the mechanical characteristics of the motor to achieve a technological process refining accuracy with significant load changes;
3. ensuring the uniformity of rotation of the motor shaft in order to ensure the cleanliness of product processing.

The choice of a cycloconverter is associated with its advantage, namely the formation of currents of a satisfactory waveform is achieved in these converters more precisely, the lower the frequency is. Therefore, it is not particularly difficult to ensure sluggish processes of metal processing.

In addition, the market appearance of DC converters (TC) of the power semiconducting reversible thyristor having linear characteristics in a wide range of voltage regulation, allows one to start the development of cycloconverters for a power supply of a hydraulic directional valve induction motor of an ultra-high-pressure hydraulic press. The possibility of completing a cycloconverter from the required number of modules of standard thyristor DC converters was shown in papers [9, 10], devoted to the analysis of technical requirements that meet specific technological processes of earthmoving machines.

When choosing the electric drive using the CCV-IM system, the following main factors were taken into account. First, as shown in Fig. 1, a drive motor of the hydraulic directional valve is an induction motor with a short-circuit rotor. Second, the main components and blocks of the electric drive control schematic diagram using the CCV-IM system were known in advance. Third, the introduction of a prepacked electric drive using the CCV-IM system in the electromechanical hydraulic press directional valve substantially extends the scope of its application.

Preliminary experimental studies in this area were conducted earlier at the request of the management of the Bishkek machine-building plant, when it was proved that it was impossible to switch to the traditionally used DC electric drive using the "Thyristor converter-motor" system. The main obstacle to this is the possibility of heavy modes of inverter triggering, which usually lead to failure of high-performance equipment [9]. The use of more modern AC electric drive systems was analyzed and the possibility of using the AC electric drive using the CCV-IM system to rotate the shaft of ultra-high-pressure hydraulic directional valve was justified.

2 Justification of the Use of a Variable-Frequency Electric Drive of the Hydraulic Directional Valve in the Servo System for the Control of the Hydraulic Press Working Member

It should be noted that operating modes of a hydraulic press during die forging or metal pressure processing, associated with the rapid movement of the working member, are similar to operating modes of the walking excavator lifting electric drive. So additional studies for the formation of static and dynamic properties of the induction electric drive are not required. However, for sluggish processes of pressure metal processing, when the movement of the working member is measured in tens or hundreds of microns per unit of time, the shaft of the hydraulic directional valve must ensure that the specified movement is refined with high accuracy with a sufficiently high movement uniformity of the working member. Therefore, the conclusion is that it is necessary to take measures to build a system to control an induction electric drive of a hydraulic directional valve. The system will provide it with high static and dynamic characteristics in the process of extruding body parts by small movements accompanied without the appearance of microcracks on the product surface and changes in the structure of the workpiece material.

For the object under consideration –a press of 30000 tons – where the positioning of the working member is controlled by a servo system, the maximum angle of mismatch $\Delta = 300°$ must be worked out by an electric drive with an error of no more than $3°$ for less than or equal to 3 s. At the same time, if we take into account that reduction gear 6 with gear ratio $i = 51$ [7] is installed between the shaft of drive motor 5 and the shaft of hydraulic directional valve 9. Then to ensure slow processes of pressure metal processing, it is necessary to set the speed of the drive motor shaft, measured by the tens or hundreds of rotations per minute.

The cycloconverter considered in this article is assembled from three bridge reversible thyristor DC converters, each of which, when connected to the stator winding of an induction motor, forms a symmetrical diagram of the power circuit. The reason for choosing bridge circuits of a thyristor converter is their high frequency properties, i.e. the frequency bandwidth of the main control parameter in a system with an optimal setting is $0 \div 45$ Hz. At the same time, the controlling effect at the input of the CCV provides the best form of changing the controlled parameter in the stator windings of an induction motor, both in static and dynamic processes.

When choosing the structure of the electric drive control system based using the CCV-IM system, preference is given to the principle of frequency-current control, since this principle is the simplest to implement for symmetrical circuits of the frequency converter. The simplicity of building a control system consists in the fact that the compensatory principle of regulating the torque by absolute slip is used, and speed control is performed by deviation. In addition, for such systems, the simplest methods of optimizing the settings of torque and speed regulators are developed. They provide a frequency-current control of the electric drive of the hydraulic directional valve with satisfactory dynamic characteristics aimed at setting the respective modes in the technological process, both in die forging and in metal processing by pressure [6]. A characteristic feature is also the

presence of functional connections that are simple and convenient for practical implementation. They ensure the maintenance of constant flux linkage of the rotor in the system
of frequency-current control of an induction short-circuited motor $\Psi_{2,\max} = const$.

The properties of the current source in the operation of the CCV are achieved by
using a one-loop control system with a proportional-integral (PI) current regulator for
each stator winding of the IM. A DC shunt integrated into the thyristor converter control
system is used as a current sensor [9, 10].

To maintain constancy $\Psi_{2,\max}$, the known relationships between the amplitude $I_{1,\max}$,
phase φ, and frequency of the stator current must be fulfilled. When the rotor flux linkage
vector Ψ_2 is oriented along the x axis of the synchronous x and y axes, to execute the
control of the torque $\Psi_2 = const$, it is necessary to ensure the formation of the stator
current vector in the form of:

$$i_1 = I_{1,\max} \cdot e^{j(\omega_{0,el} t + \varphi)} \tag{1}$$

Here

$$I_{1,\max} = \frac{\Psi_{2,\max}}{L_{12}} \sqrt{1 + \frac{L_2^2}{R_2'^2} \omega_{0,el.n.}^2 s_a^2}, \tag{2}$$

$$\varphi_1 = arctg(L_2 \omega_{0,el.n.} s_a / R_2') \tag{3}$$

where $\omega_{0,el.n.}$ – nominal electric speed of the stator field; s_a – absolute slip; L_2, L_{12} –
inductance of the rotor and the mutual inductance between the stator and rotor windings
respectively; R_2' – resistance of the rotor brought to the stator circuit.

Expressions (1)–(3) are written on the basis of a vector diagram (Fig. 2), which shows
that to maintain the flow $\Psi_2 = const$ in real time, it is necessary, as in any vector control,
to change not only the amplitude $I_{1,\max}$, but also the phase φ_1 of the stator current vector
$\dot{I}_1$.

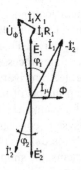

Fig. 2. Induction motorvector diagram.

When forming the stator current according to the formula (1), the mechanical
characteristic of an induction motor has the following form [11]:

$$M = \beta(\omega_0 - \omega) = \beta \cdot s_a \cdot \omega_{0,n}. \tag{4}$$

where $\beta = \frac{\Psi_{2,max}^2}{R_2'} p_p^2$.

From expression (4), it follows that theoretically, for $\Psi_{2,max} = $ const, the torque control loop is inertia-free and the motor instantly reacts to a change in speed by changing the torque.

However, in the system under consideration, the constancy of the rotor flux linkage is provided by a compensatory way, and, consequently, the accuracy of maintaining the constancy of $\Psi_{2,max}$ is limited. The possibility of minor changes in the motor flow determines the electromagnetic inertia, which can be taken into account by a small T_e constant in the following expression:

$$(1 + T_\vartheta p)M = \beta \cdot s_a \cdot \omega_{0,n}. \tag{5}$$

The accuracy of motor torque control according to expression (5) is determined by an accurate measurement of the speed ω, the values of which are determined by the DC tachogenerator built into the induction motor. In addition, it is necessary to emphasize other roles performed by a signal proportional to the speed of the IM shaft. One of them is the setting of the stator current frequency based on the following ratio:

$$\omega_0 = \omega \cdot p_p + \omega_{0,el} \cdot s_a \tag{6}$$

and control of the speed of the electric drive by deviation.

The formation of the main harmonics of the IM phase current by a frequency converter (1) imitating the form of the driving voltage is a combined control system. Here the current control loop of the stator winding is represented by a transfer function of PI current controller W_{cc}, a TC thyristor converter, covered by positive feedforward with transfer function W_{ce} and a mathematical model of a phase winding (Fig. 3).

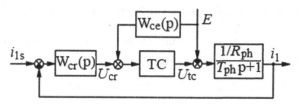

Fig. 3. Block diagram of the combined ADS of IM phase currents.

In case of a slowly changing motor EMF, the PI-current controllers built in the thyristor converter, configured to the technical optimum, allow providing a satisfactory quality of the current waveform at the required response speed. However, in a fairly wide range of operating frequencies, the quality of phase current regulation is affected by the motor's counter-EMF, the amplitude and frequency of which grow in proportion to the frequency of the output voltage of the CCV. The perturbing effect of the counter-EMF results in amplitude and phase distortions in the waveform of the output current of the CCV, and compensation of this disturbance is necessary for an acceptable quality of current control.

The waveforms in Fig. 4 show the principle of compensation for a perturbing effect, taken in the operating frequency range, both with low load and at full load. Considering these oscillograph records, we can state that there is a significant phase shift between the current setting signal and the actual phase current. This circumstance leads to a tendency of unstable operation of the system with characteristic knocks, since the distortion of the current waveform causes additional disturbances and generates noise and vibration of the current.

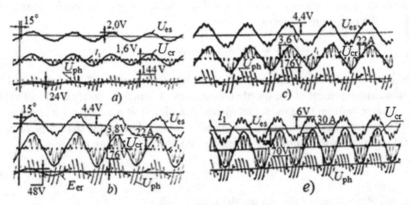

Fig. 4. Oscillograph records of currents and voltages illustrating the principle of compensation of IM rotation EMF: a), b) – without compensation; c), e) – with compensation; a), c) $M_{heat} = 0.4\,M_{nom}$; b), e) $M_{heat} = M_{nom}$.

The motor rotation EMF effect, according to the block diagram shown in Fig. 3, is compensated by using a feedforward on EMF of the motor applied to the input of the TC amplifier. An additional winding wound on the poles of an induction motor can serve as an EMF sensor. In the absence of special windings for measuring the EMF of rotation, it is advisable to use an indirect method based on solving the equation of electrical equilibrium in the stator winding:

$$E_{rot} = U_{1,ph} - i_1 R_1 - L_1 di_1/dt \tag{7}$$

where E_{rot} – induced EMF in the stator phases (Fig. 4, b); $U_{1,ph}$ – voltage at the output of the CCV; i_1 – stator current; R_1, L_1 – active resistance and inductance of the stator phase scattering.

To solve Eq. (7), it is necessary to use analog circuitry based on modern high-speed operational amplifiers (OAs). The reason for this is that the current regulator is implemented using an OA, the input of which is a signal proportional to the stator current, measured using a DC shunt. Therefore, the same signal can be used to calculate the EMF of the IM rotation. In addition, the environment where the operational amplifiers that implement a current controller and EMF sensor operate is associated with high-frequency interference in high-voltage circuits. To measure the phase voltage for each stator winding, voltage dividers are used, which are constructed using fixed-value resistors.

The principle schematic diagram of the EMF sensor implementing Eq. (7) is shown in Fig. 5. In this schematic diagram, the signal proportional to the stator current is differentiated and summed at the input of the OA2 with the component of the voltage drop on the active resistance of the stator winding. The resulting sum of voltages is fed to the input of the OA3, where it is added to the phase voltage at the output of the CCV.

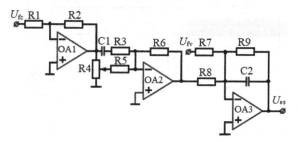

Fig. 5. Schematic diagram of the EMF sensor.

The effect of EMF of IM rotation according to Fig. 3 is compensated by introducing feedforward according to the EMF of the motor with the transfer function:

$$W_{ce} = \frac{K_{ce}}{T_{ph}p + 1},$$

where by selecting the amplification constant of the K_{ce} in the channel for regulating the perturbing effect, it is possible to achieve a practical coincidental occurrence of the main harmonics of the current with the set one.

The presence of an aperiodic link in the OA3 feedback circuit is determined by the need to filter high-frequency pulsations of the motor EMF sensor that adversely affect the operation of the phase current control loop. It is experimentally established that a satisfactory quality of the generated current is obtained if the time constant T_{ph} is within $0.1 \div 1$ ms.

Figure 6 presents a functional schematic diagram that shows the main elements by which the properties of the current source are set for the cycloconverter over the entire frequency control range of the IM stator windings. The measures taken to compensate for the perturbing effects of rotation EMF made it possible to generate phase currents of the induction motor in accordance with expression (1) with an accuracy sufficient to ensure the technological processes of the hydraulic press.

If the CCV in Fig. 6 is represented as a multi–pole element of automation, then the parameters are the voltage and current varying according to the sinusoidal law. The voltage of the current setting U_{crA}, U_{crB}, U_{crC} is input parameters; the phase currents of the induction motor I_{ph1A}, I_{ph1B}, I_{ph1C} are output parameters.

Thus, it is necessary to create a three-phase symmetric sinusoidal signal generator (SSG) with an operating frequency of generating $0 \div 45$ Hz. Moreover, according to expressions (1)–(3), in order to maintain the rotor flow IM$\Psi_2 =$ const, it is necessary for the SSG to generate three-phase signals, in which both the frequency and amplitude, and the phase of the induction motor current setting voltages are regulated from the input

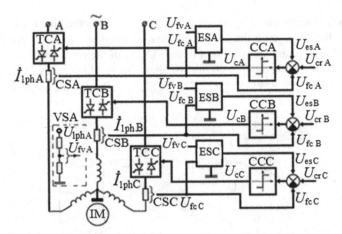

Fig. 6. Functional schematic diagram of a cycloconverter that has the properties of a current source: TC – reversible thyristor converter of stator windings phases; CS – current sensors of IM phases; ES – EMF sensors; VS– voltage sensor of phase A; CC– PI-current controllers of IM phases.

signals. Therefore, the voltage equation that sets the current, for example, for phase A, is written as:

$$u_{crA} = U_{crm} \sin(\omega_0 t + \varphi_a) \tag{8}$$

where U_{crm} – amplitude value; ω_0 – angular frequency of the IM variable vectors rotating synchronously with the stator field; φ_a – phase of the generated voltage.

In order to determine the relationship between the amplitude and phase of the generated voltage, let us decompose Eq. (8) into two components:

$$u_{crA} = U'_m \sin \omega t + U_C \cos \omega t \tag{9}$$

Considering the vector diagram in Fig. 2, we can conclude that if we use a term that changes according to the cosinusoidal law with a constantly set amplitude value U_c to set the magnetizing current in Eq. (9), then the constancy of the rotor flux linkage is ensured $\Psi_2 = \text{const}$. In this case, the component that changes according to the sinusoidal law with the amplitude U'_m is a regulated voltage, which compensates for the "drawdown" of the speed caused by an increase in the load on the motor shaft.

With these notations, the expressions for calculating the amplitude modulus U_{crm} and a phase φ_a have the following form:

$$U_{crm} = \sqrt{(U'_m)^2 + (U_C)^2}, \quad \varphi_a = arctg(U_c/U'_m) \tag{10}$$

The theoretical calculations on the equations established that the voltage of the current setting and its amplitude are proportional $u_{crA} = U'_m$. However, it is difficult to establish an unambiguous proportionality between these variables in theory. So in practice a different relationship $U'_m = kU_{tr}$ is used, written based on the principle of operation of an automated electric drive, according to which a decrease in the stator current I_1 unambiguously causes a decrease in the motor torque. As a result, there will be a speed drawdown, in response to which the speed controller will increase the torque setting. Increasing the motor torque is possible only by increasing the current amplitude, etc.

The value of the coefficient k, which establishes the relationship between these voltages, is unknown in advance, but during the adjustment process, it is experimentally possible to set an acceptable value for a number of speeds that provide a step-by-step cycle of the slide (Fig. 7).

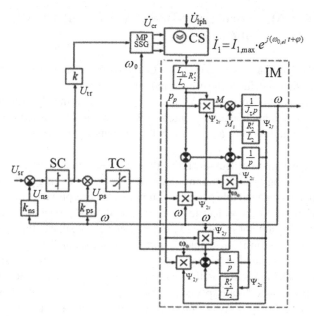

Fig. 7. Block diagram of the control of the hydraulic directional valve electric drive using the *CCV*-IM system: CS – current source supplying the IM winding; IM – two-phase model of the generalized machine when powered by the CS; SC– PI-speed controller; TC– proportional torque controller; MPSSG– three-phase microprocessor-based sinusoidal signal generator.

Figure 7 shows a functional schematic diagram for controlling the electric drive of a hydraulic directional valve using the CCV-IM system, which indicates the location of a microprocessor-based sinusoidal signal generator (MPSSG).

In this schematic diagram, an induction motor is represented as a block diagram based on the equations of a generalized two-phase machine powered by a current source [11–13]. An internal torque control loop optimized for the modular optimum is formed with the help of the TC torque controller, and the speed controller (SC) of the proportional-integral structure forms an external loop, a speed control loop configured for the modular optimum [6, 14]. The output signals of the TC and SC regulators are the input signals of the MPSSG, respectively: the output signal of the TC is a frequency setting ω_0; the output signal of the SC is the U_{crm} amplitude setting generated by CS for the phase currents of the IM windings. MPSSG is assigned the role of generating control signals for setting currents in the IM phases according to the embedded algorithm of equations of the following form:

$$\begin{cases} u_{crA} = U'_m \sin \omega_0 t + U_C \cos \omega_0 t, \\ u_{crB} = U'_m \sin(\omega_0 t + 120°) + U_C \cos(\omega_0 t + 120°), \\ u_{crC} = U'_m \sin(\omega_0 t - 120°) + U_C \cos(\omega_0 t - 120°), \end{cases} \tag{11}$$

where $U'_m = kU_{tr}$.

3 Design of a Microprocessor-Based Version of Creating a Sinusoidal Signal Generator

The choice of the microprocessor-based version of creating a sinusoidal signal generator is justified by the fact that due to the use of modern high-speed microprocessor systems, it became possible to switch to a software method for implementing algorithms for solving Eqs. (11). That is, this method allows entering the structure of a three-phase sinusoidal signal generator in the form of a program without changing the hardware and circuitry framework.

The advantage of microprocessors is that they have built-in peripherals, which greatly facilitate programming or reprogramming in the process of writing and debugging the software product, using available personal computers for this purpose. In operating conditions of a variable-frequency electric drive, the use of microprocessor systems is much preferable, since they are not sensitive to external interference due to the fact that, firstly, the processed information is presented in discrete form; secondly, measures to ensure noise immunity are planned in advance for such devices. In addition, a small number of elements (modules) and connections between them ensures high reliability of the generator. It is possible to create special diagnostic programs and significantly facilitate the troubleshooting in the software product and ensure the adaptation of continuous generation of harmonic signals.

When designing any control microprocessor devices in an automated electric drive system, the mandatory implementation stages must be followed, according to which equations describing the flow of physical processes are compiled at the beginning, and based on them; the control devices based on the electric drive control features are built. Next, a functional schematic diagram of the mathematical model for solving equations is compiled. Then a block diagram is compiled for writing a program using a high-level programming language. The next step is to select the element base that provides for the implementation of each cell, node, block, or device. Next, a schematic diagram is drawn up, the printed circuit board is developed, and the printed circuit board is installed. The final stage is adjustment; all parameters of control elements are brought to the required level during it.

Figure 8 shows the functional schematic diagram of the model of the developed microprocessor generator, which shows the main blocks that implement the system of Eqs. (11).

In this schematic diagram, the variable input signals are: a signal for setting the frequency ω_0, taken from the output of the TC controller; a signal for setting the amplitude

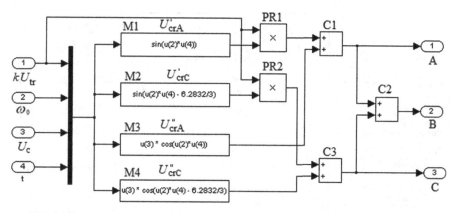

Fig. 8. Functional block diagram of a three-phase sinusoidal signal generator model.

of the kU_{tr}, taken from the output of the SC controller. A constant input signal is U_c, which is set once before starting and is maintained at a constant level during the operation of the MCSSG, since this parameter sets the magnetizing current to the IM. The parameter t is the working variable for the M1, M2, M3, M4 calculation blocks, in which the harmonically varying terms of the 1st and 3rd equations of the system (11) are calculated. Using PR1 and PR2 multiplication blocks, the results obtained from M1 and M2 outputs are multiplied with the amplitude values of sinusoidal signals, the level of which is determined by the speed of the motor. Adders C1 and C2 are used to sum the terms of the 1st and 3rd equations of the system (11). Thus, using such blocks (Fig. 8), signals for setting the currents of phases A and C of the induction motor are generated, and in order to get a signal for setting the current of phase B, a simple algebraic summation using the C3 adder is sufficient.

The algorithm of the microcontroller SSG software is represented by the block diagram in Fig. 9, which shows that the operation of the microcontroller begins with the preparation of input ports for performing measurement and calculation operations, i.e. first initialization of all the necessary peripherals that the controller will work with (ports, timers, PWM, etc.). Next, the program initializes the timers, which consists in the fact that the timers in the discrete generator mode form samples, relative to which the trigonometric functions are converted into step functions, and the smooth component of these functions is closer to the original, the shorter the quantization time.

In addition, the accuracy of measuring input signals using the ADC built in the microcontroller is also determined by the quantization time. The specified sampling rate of 66 kHz is sufficient to provide continuous generation of controlling effects on the ADC and PWM for converting input analog signals into the digital code, as well as calculating the output signal as a step function with high accuracy.

The algorithm shown in Fig. 9 shows that the program execution is divided into two threads. In the first thread, the input variable signals of the SSG are measured using an ADC and a timer, and the magnetizing current of the induction motor is set as a constant, previously obtained as a result of calculation, or its value will be determined

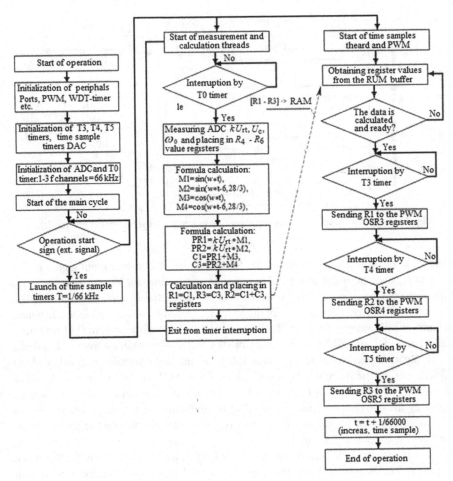

Fig. 9. Block diagram of a sinusoidal signal generator program.

experimentally during setup. The obtained measurement results and constant data are then placed in the storage registers R1-R3 by the program.

According to the block diagram (Fig. 9), the next action of the program is to calculate trigonometric functions in blocks M1-M4 according to the time and frequency with preset initial conditions and constants. The results of calculations of blocks M1-M4 are further processed using the product blocks PR1, PR2 and summation blocks C1–C3 according to the algorithm shown in the functional diagram in Fig. 8, and are grouped into data arrays for transmission to the storage registers R1-R3. As the dotted line with an arrow in Fig. 9 shows, the results of measurements and calculations stored in registers R1-R3 are then transferred to the buffer to the random access memory (RAM) to synchronize this data with the 2nd thread algorithm.

Execution of the second thread of the algorithm can only begin after the first thread operations are performed, and data will be transferred to RAM from registers R1-R3. In the direction of the second thread, the sinusoidal voltages U_{csA}, U_{csB}, and U_{csC}

calculated in the digital code are ready for digital-to-analog conversion using the built-in DAC controlled by PWM with further transmission to the corresponding OCR3-OCR5 registers. In this thread, the PWM sampling rate is a third of the ADC measurement sampling rate and is equal to 22 kHz. The end of the 2nd thread is completed by the action of incrementing the samples according to time t.

The operation of the microprocessor, whose algorithm is described above, will be repeated until the WDT timer is triggered, which protects the program from hardware failure or "freezing". Figure 10 shows a schematic diagram of a three-phase sinusoidal signal generator with an indication of the selected element framework, which implements the microcontroller operation algorithm described above.

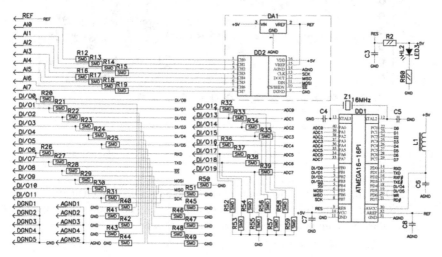

Fig. 10. Schematic diagram of a microprocessor-based sinusoidal signal generator.

The element framework that allows implementing a digital sinusoidal signal generator is an 8-bit high-performance AVR microcontroller *Atmega* 16-16 P1 with low power consumption and a speed of up to 16 million operations per second [15, 16]. The microcontroller has a progressive RISC architecture with 130 high-performance commands, with most commands executed in a single clock cycle by 32×8-bit general-purpose working registers. The D0-D7 bits of the microcontroller (8 bits) are converted to an analog OUT signal. The operating voltage of the microcontroller is 4.5–5.5 V. These chips are produced in large quantities, have a relatively low cost, operate at a wide range of ambient temperature changes. They also have an accessible programming and reprogramming system, and have a high degree of protection against interference, so they do not require special conditions during operation.

In the schematic diagram, the DA1 element is a conventional dual operational amplifier (OA). It performs the role of matching the levels of the resistor DAC R-2R. The DD2 element is an external backup ADC that is provided for more complex measurements with multiple variables. The DD2 element does not participate in the operation of the sinusoidal signal generator. Figure 11 shows a printed circuit board as a layout diagram to indicate the completion of the development of a control device – a sinusoidal signal generator.

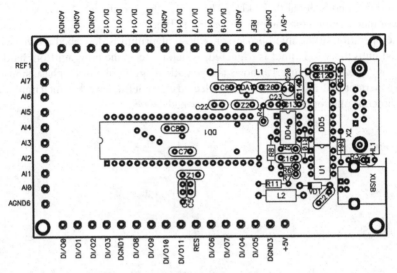

Fig. 11. Layout diagram of elements a microprocessor-based sinusoidal signal generator.

The DD1 microprocessor is the main element, and 3 symmetrical channels based on resistor DACs built on the basis of the DA1 OA are auxiliary elements. Other elements specified on the board are not involved in the generator operation.

4   Conclusions

The following conclusions were made in the course of the featured study.

1. It has been established that the operation of the CCV in the current source mode, compensated from the effect of the EMF of induction motor rotation, can ensure uniform rotation at low rpm of the motor shaft in the presence of a sinusoidal signal generator.
2. The use of microprocessor-based sinusoidal signal generator to control the CCV-IM, as well as the corresponding placement of the hydraulic directional valve cam shaft profile, provides the ability to configure a hydraulic press to different modes of processing (including the draw-forming of body parts of large dimensions) in accordance with the required products processing technology.
3. For the formation of 3-phase sinusoidal signals, the most appropriate schematic diagram is based on a microcontroller with 3 8-bit ports and 3 resistor DACs with matching OAs on DA1.
4. Sinusoidally varying voltages for setting the phase currents of the stator, shown in the oscillograph records in Fig. 4, confirm the fact that the MPSSG generates sinusoidal signals with a high degree of accuracy.

5. The advantage of the developed microprocessor-based sinusoidal signal generator is its versatility, i.e. it can be used both for controlling variable-frequency electric drives, and for configuring equipment with a setting for the input of a harmonic controlling effect, for example, for removing frequency characteristics, since the frequency of generation can vary widely.

References

1. Ovchinnikov, A.G.: Progressive Technological Processes of Cold Forging. Mashinostroenie, Moscow, 184 (2001)
2. Ovchinnikov, A.G.: Fundamentals of the Theory of Impact Extrusion on Presses. Mashinostroenie, Moscow, 200 (2002)
3. Khokhlov, V.A., Prokofiev, V.N., Borisova, N.A., et al.: Electrohydraulic Servo Systems. Mashinostroenie, Moscow, 430 (1981)
4. Gomynin, N.S.: Fundamentals of Hydraulic Servodrive. Oborongiz, Moscow, 430 (1981)
5. Popkov, S.L.: Servo Systems. HigherSchool, Moscow, 304 (1975)
6. Kadyrov, I., Karaeva, N.S., Andarbekov, Z.A., et al.: Principles of electric drive remote control contruction using the CCV-IM system for a hydraulic directional valve of a 30,000 tons' press. KSTU News **3**(51), 95–105 (2019)
7. Mikheev, V.A., Yam, V.M., Polyakov, B. I.: Modernization of Hydraulic Press Equipment. Mashgiz, Moscow, 251 (1951)
8. Bochkarev, I. V., Kadyrov, I.Sh.: Microprocessor control device using the "Cycloconverter – IM" system of the excavator electric drive. News of higher educational institutions. Electromechanics **5**, 25–30 (2007)
9. Kadyrov, I.Sh.: Study of dynamic processes in the gears of the excavator-dragline rotation mechanism when the supply voltage is suddenly switched off. Mining Equipment and Electromechanics **6**, 29–31 (2008)
10. Bochkarev, I.V., Kadyrov, ISh: Optimization of parameters of the control device for the excavators' electromechanical system. Electr. Technol. **2**, 2–7 (2009)
11. Kliuchev, V. I.: Theory of Electric Drive. Energoatomizdat, Moscow (1985)
12. Robertson, S.D.T., Hebbar, K.M.: A digital model for three phase induction machines. IEEE Trans. Power Apparatus Syst. **88**(11), 1624–1634 (1969)
13. Kadyrov, I.Sh.: Management of Technical Systems: A Textbook for Universities. KGTU, Bishkek (2014)
14. Bessekerskiy, V.A., Popov, E.P.: Theory of Automatic Control Systems. Nauka, Moscow (1972)
15. 8-bit AVR Microcontroller with 8 K Bytes In-System Programmable Flash. ATmega8535. Product Datasheet. Atmel Corp. (2003)
16. Choudhury, S.: Average Current Mode Controlled Power Factor Correction Converter using TMS320LF2407. Application Note SPRA902A. Texas Instruments (2003)

Program Model of the Interacting Adaptive Traffic Control System

Timofei Radionov[1] , Anton Mikhalev[1] , Vladislav Kukartsev[1,2] ,
and Vadim Tynchenko[1,2,3(✉)]

[1] Siberian Federal University, Svobodnyy Ave 82A, Krasnoyarsk 660041, Russia
vadimond@mail.ru
[2] Reshetnev Siberian State University of Science and Technology, Gazety Krasnoyarsky
Rabochy Ave. 31, Krasnoyarsk 660014, Russia
[3] Marine Hydrophysical Institute of Russian Academy of Sciences, Kapitanskaya Street 2,
Sevastopol 299011, Russia

Abstract. The article is devoted to the issue of traffic management. A review
of existing software solutions for simulating traffic situations has been carried
out, and several approaches to optimizing the operation of traffic lights have been
described. The algorithms and features of the functioning of modern automated
traffic control systems are described. A developed software model, its interface,
functionality, and a device for studying various algorithms for traffic lights regulation at two intersections with four-way intersection are considered. A traffic
light signal regulation algorithm developed by the authors is presented, which is
a modification of the existing adaptive regulation algorithm. The article presents
the traffic light regulation algorithm developed by the authors, which is a modification of the existing adaptive regulation algorithm. The structure and operation
of this algorithm are described mathematically. The analysis and comparison of
the presented algorithm with the existing control algorithms in terms of efficiency
and time are carried out. Based on the results obtained, conclusions were made
about the possibility of introducing the author's algorithm considered in the article
into a real process.

Keywords: Transport · Intersection · Congestion · Traffic light control system ·
Roadway traffic · Automated traffic control

1 Introduction

Many cities have difficulty with traffic. Therefore, cities need innovative ways to deal
with heavy traffic situations. Traffic lights have the most significant impact on traffic,
but traffic lights drive traffic. Improving traffic light performance can have a significant
impact on the traffic situation in the city as a whole.

© Springer Nature Switzerland AG 2020
V. Jordan et al. (Eds.): HPCST 2020, CCIS 1304, pp. 218–234, 2020.
https://doi.org/10.1007/978-3-030-66895-2_14

Currently, most traffic lights in the Russian Federation operate according to pre-configured signal change programs. This type of traffic light works as a simple mechanism for changing signals with a configured timer for each of them. Further in the article, such traffic lights will be referred to as standard traffic lights. There are also modifications to these traffic lights. However, their essence is unchanged. Namely, traffic lights are not flexible enough about the road situation, and they have a standard traffic light control system (STLCS).

In world practice, an adaptive traffic light control algorithm (ATLCA) is often used. The peculiarity of the adaptive traffic light control algorithm is that it adjusts its signal-switching mode to the traffic flow regulated by it using additional devices. This type of traffic light is still not widely used in Russia. However, this implementation of this type of traffic light is a matter of time.

This article proposes its control algorithm, called the Interactive Adaptive Traffic Light Control System (IATCS), which is a modification of ATLCA. This algorithm implements the interaction between traffic lights and takes into account the distance between the corresponding intersections. The signal is adjusted to prevent vehicles from fully covering the distance between the regulated intersections. This circumstance often leads to traffic congestion and requires permission. In order to study IATCS and compare with the algorithms considered in this article (STLCS and ATLCA), a software model was developed. Comparison of the algorithms in terms of efficiency and time costs was produced, taking into account the functioning of the algorithms when operating on the same simulated traffic situations.

2   Literature Review

During the study, we reviewed existing approaches to optimizing traffic light performance and the simulation programs used for this. One of the approaches to improve the organization of traffic lights is a traffic light control system based on a multi-agent model. A feature of the multi-agent model is the control of the green offset value using only local information about the situation. The formation of a "green wave" occurs through the coordination of each crossing agent, which ensures smooth and efficient movement in a certain direction [1]. Of particular interest is a model that uses infrared proximity sensors and a central microcontroller, as well as vehicle length measurements. These tools allow implementing an intelligent traffic monitoring system, which should help solve the problem of traffic jams [2]. There is a model that uses the designation model optimization process, with the additional use of modern video processing techniques instead of infrared sensors. Such a system manages standard four-way intersections using three parameters, namely, the average vehicle speed, the occupancy rate of the intersection with vehicles and the emergency timer. Video from traffic cameras using computer vision algorithms determines these three parameters. The Adaptive Traffic Management (ATMC) decision process was designed to adapt traffic parameters to pre-defined priorities. The proposed ATMC was verified using four synchronized test videos, which demonstrated full adaptation to the transport stream [3].

Another approach is based on traffic data and the congestion principle by introducing a congestion factor. This coefficient can be applied to assess the state of traffic congestion in real-time along the segment and to predict the sub-critical state of traffic jams. This approach made it possible to optimize the synchronization phase of traffic light control and prevent traffic congestion. Traffic monitoring models based on the Mobile Century dataset and analysis of traffic light management efficiency on the VISSIM platform are presented in [4]. The use of modern methods of neural networks and fuzzy logic has led to successes in improving the performance of traffic lights. This approach is interesting because there is no need to use mathematical models when modeling traffic lights. Vehicle detection and measurement of their speed are based on the application of acoustic and radar sensors. As a result of these inputs, the traffic light timer was adjusted [5]. Consideration of the existing models of traffic light operation leads to the conclusion that at the current time, the problem of optimization of regulation of traffic signals is relevant and the ways to solve it are various. The authors propose a fundamentally new IATCS algorithm for solving this problem, which is aimed at reducing the occurrence of congestion at a controlled intersection. This approach to solving the problem of regulation of traffic flows has not yet been considered in scientific articles devoted to this problem.

3 Traffic Light Algorithms

Let us consider a traffic light device. The traffic light includes a sequence of light sources and a control unit. In the simplest case, the control unit is an electromechanical device that switches signals at a predetermined interval in a cyclic sequence according to the STLCS algorithm (Fig. 1). In more difficult cases, the traffic light changes the signal based on external influences, such as pressing a button by a pedestrian or stopping the vehicle on a weight sensor.

Control units of modern traffic light models are based on microprocessor systems. Connecting a traffic light to a computer network allows the operator to change signal-switching intervals based on the traffic situation. Also, the use of control units based on microprocessor systems makes it easier to change the traffic light algorithm by changing the program [6].

Let us consider the operation of adaptive traffic lights using the example of an algorithm that skips queues formed during the period of the prohibiting signal. There are two types of detectors at the intersection: D1 and D2 (Fig. 2). D1 detectors are located at the entrance to the intersection. D1 detectors read the license plate numbers of vehicles at the moment of passing through them and enter the numbers into memory. D2 detectors are required to keep track of vehicles leaving the intersection. Therefore, D2 detectors are located directly at the border of the intersection. The interactions between detectors and traffic lights are as follows.

An identification number is read on all vehicles in the system (for example, state registration number, RFID identification code). During the operation of the red phase of the traffic light, D1 detectors read the numbers of the vehicles passing through them and enter them into the list. The moment the green light comes on, detector D1 stops recording license plates, and detector D2 begins to remove license plates from the list

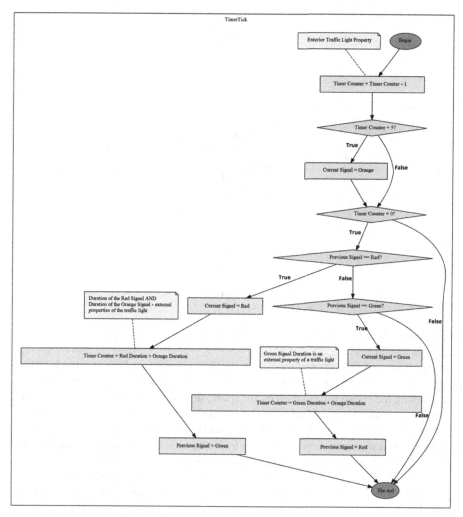

Fig. 1. Block diagram of the STLCS algorithm.

of vehicles passing through it. At the moment when the last of the vehicles pass through detector D2, and the list becomes empty. The traffic light switches to the red phase and goes to the beginning of the cycle. This rule is valid for both roads in question.

In the presented case, ATLCA is an algorithm that provides for the redistribution of the duration of the phases within the cycle based on the analysis of the current phase coefficients in conflicting directions. This algorithm works on the principle of changing the duration of the phases of a fixed traffic light, depending on the intensity of incoming traffic flows (Fig. 3, Table 1). This algorithm does not use blocks marked with "! IATCS ". In this case, a certain number of traffic light cycles determine the intensity. It is enough to use one registering detector. During a certain period, the intensities are counted.

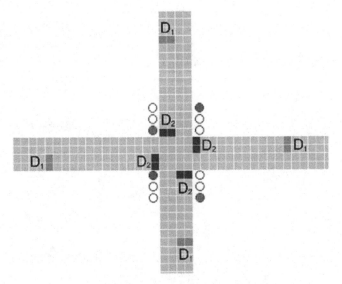

Fig. 2. Layout of detectors.

Depending on the intensity change, the duration of the traffic light phases changes by a particular value [7–9].

The authors of this article propose the IATCS algorithm, which is a modification of the ATLCA algorithm described above. The modification of the algorithm consists in expanding the stored information by the traffic flow detectors. This algorithm makes it possible to estimate the duration of the green signal phase taking into account this additional information, and not only by the number of vehicles. Traffic lights and the exchange of this information between other traffic lights at adjacent intersections the distance between intersections control additional information. The IATCS algorithm works similarly to the ATLCA algorithm described above. However, there are fundamental differences. A significant difference lies in the step of estimating the duration of the green signal phase (Fig. 4) and the use of blocks of the ATLCA algorithm, including those marked "! IATCS" (Fig. 3).

Application of the IATCS algorithm for traffic light regulation:

1. the duration of the green signal phase is estimated along the auto stream in front of the detector;
2. there is a collection of information from the nearest traffic lights, to which the flow will be directed, about the length of the road between the current and the nearest intersection and the number of vehicles (length of the flow) on this road;
3. using the information received at step 2, the green phase duration is adjusted so that the current traffic light passes as many vehicles as the road section next to the intersection can afford.

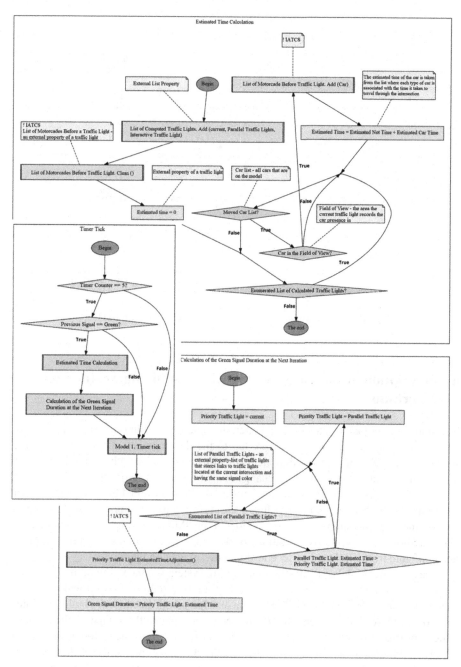

Fig. 3. General block diagram of the ATLCA and IATCS algorithms.

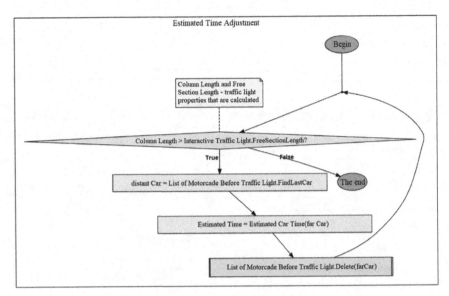

Fig. 4. Block diagram of the estimated time correction in IATCS.

4   Description of the Software Model for Studying Traffic Light Algorithms

The study set the task of checking the operation and researching the traffic light algorithm. The implementation of this task led to the development of our own software model. A feature of the proposed model is that this model allows changing the existing algorithms of traffic lights, making changes to the simulated traffic flow measurement sensor. This model can also work with several algorithms of traffic lights to compare their performance by observing and recording useful information. The program interface is a multi-window application. The main window of the "Model settings menu" application (Fig. 5) allows setting the basic settings of the model: build routes for vehicles, place and configure traffic lights, display additional useful data on the model.

The software model of the traffic light control system is a background image, which can be a satellite image of a real area or a diagram of a road section on which objects specified by the user are located. The system also has a built-in editor for building routes (Fig. 6). In the editor, the user creates routes (according to the image of the model), which are roads for vehicles. There is a function for saving routes. Routes are stored locally in a file as a set of lines. It is noteworthy that the coincidence of the points of two segments leads to their union. This function adds convenience and speeds up the process of creating paths for vehicles.

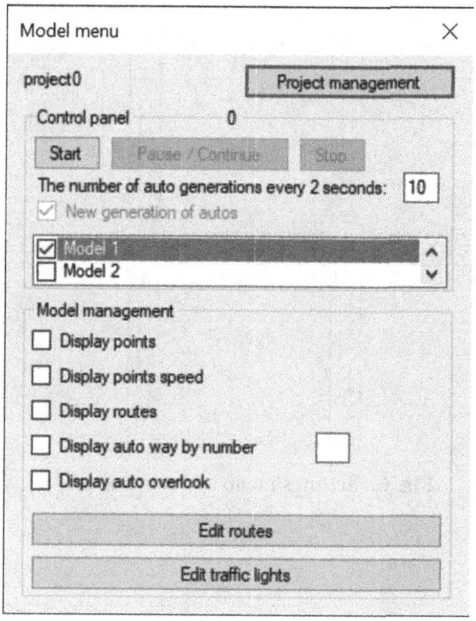

Fig. 5. Program model settings menu.

Important in the program is the traffic light editor (Fig. 7), which allows editing the initial traffic light settings on the model. The program is configured so that changes in the traffic light timer are automatically synchronized with other light norms. As a result, there is an adjustment of traffic signals, which are directly related and should not have contradictions. This block uses a partially adaptive principle of signal regulation. The partially adaptive principle of signal regulation can be used when editing signals and when executing a program, for example, for ATLCA.

The ability of the software model to simulate the movement of vehicles is of great importance for the development of the software model (Fig. 7). The movement algorithm is completely developed from scratch. Each vehicles object acts according to the established rules, has its field of view and can interact with other objects of the model.

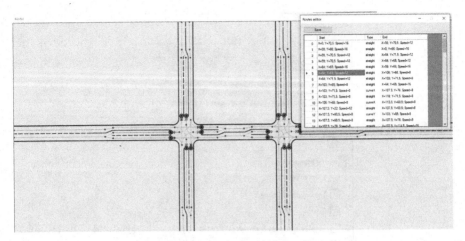

Fig. 6. Settings menu and route editor.

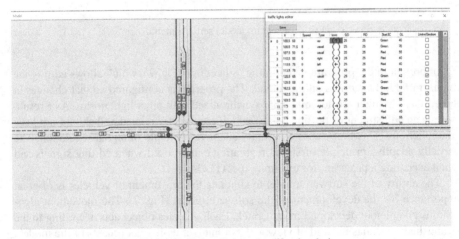

Fig. 7. Traffic light editor and traffic simulation.

5 Mathematical Apparatus

Each considered traffic light has other traffic lights of the intersection dependent on it (Fig. 8). Traffic lights are located on routes that cross the route with the considered one; they are perpendicular traffic lights. Traffic lights can be co-directional both in one direction and in the opposite direction, provided that the route controlled by them is directly or indirectly (through dependent) connected with the traffic light in question (parallel traffic lights). Also, for the IATCS algorithm, each traffic light can have a link to another traffic light on a subsequent limited road section (interacting traffic light).

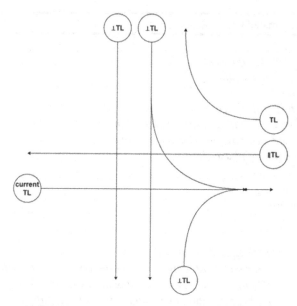

Fig. 8. Dependence of traffic lights at an intersection.

When the yellow traffic light is on, the process of calculating the duration of the green traffic light phase is started. The duration of the phase is about 5 s. For STLCS, these calculations are absent since the traffic signal state changes immediately (Fig. 1). For ATLCA and IATCS, the calculation takes place in two stages:

1. EstimatedTimeCalculation – calculation of the required green signal burning time for the convoy (calculations are performed separately for each group of interconnected traffic lights, as well as for the interacting traffic light – in the case of the IATCS algorithm);
2. CalculateGD (Formula 2) – calculation of the green signal duration, selection of the priority signal, "equalization" of the green signal duration for the remaining traffic lights (calculations are performed separately for each group of interconnected traffic lights).

For the IATCS algorithm, after determining the priority traffic light (CalculateGD stage), the EstimatedTimeCorrection stage is performed. The EstimatedTimeCorrection stage is a correction to the calculated green signal time based on information from the interconnected traffic light about the presence of a free area on the subsequent road section (Fig. 9).

Pseudocode EstimatedTimeCalculation

Input

TL_{cur} *// traffic light in question*
$\perp TL$ *// perpendicular traffic light*
$\| TL$ *// parallel traffic light*
ITL *// interacting traffic light*
auto // set of vehicle waiting for the enable signal
$time_{min}$ *// set of vehicle waiting for the enable signal*
$time_{shift}$ *// amount of phase shift*

Initialize

$TL_{cur}.ET = time_{min}$
$\forall \| TL.ET = time_{min}$
$ITL.ET = time_{min}$

begin

    for TL in $\{TL_{cur}, \forall \| TL, ITL\}$
        // for each traffic light, depending on the number of vehicles,
        // being in its field of view, the minimum time is adjusted
        // phase duration by the amount of phase shift
        if $auto \subset OL \rightarrow TL.ET = TL.ET + time_{shift}$

    end

end

Pseudocode CalculateGD

Input

TL_{cur} *// traffic light in question*
$\| TL$ *// parallel traffic light*

Initialize

$PTL = TL_{cur}$ *// set the priority traffic light*

begin

    for $\forall \| TL$
        if $PTL.ET < \| TL.ET \rightarrow PTL = \| TL$

    end

    EstimatedTimeCorrection // performed only for IATCS
    $TL_{cur}.GD = PTL.ET$ *// set the duration of the green signal*

    for $\forall \| TL$
        $\| TL.GD = PTL.ET$ *// set the duration of the green signal*
    end

end

Pseudocode EstimatedTimeCorrection

Input
ITL // traffic light with which the current one interacts
FS // length of the free road section
AL // length of the flow of vehicles in front of the traffic light (in its field of view)
OL // traffic light scope
TL_{cur} // traffic light in question
$time_{shift}$ // amount of phase shift
Initialize
$ITL. FS = ITL. OL - ITL. AL$
begin
$$\textbf{while } TL_{cur}. AL > ITL. FS \rightarrow TL_{cur}. ET = TL_{cur}. ET - time_{shift}$$
end

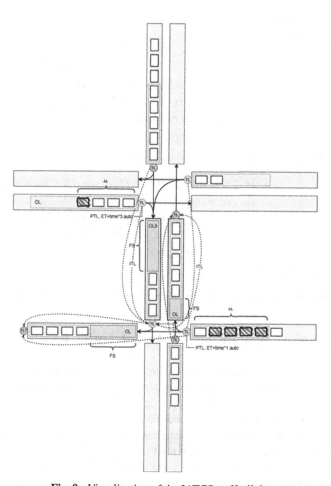

Fig. 9. Visualization of the IATCS traffic lights.

Synchronization starts after calculations and received values. Synchronization is a check of the values of the traffic light signals' duration interconnected at the intersection and the selection of those values that will avoid possible logical inconsistencies in terms of the correct operation of traffic lights at the intersections.

Pseudocode SignalTiming

Input
SC // signal color
TL_{cur} *// traffic light in question*
$\perp TL$ *// perpendicular traffic light*
$\parallel TL$ *// parallel traffic light*
SD // duration of signal burning
GD // green signal duration
RD // red signal duration
begin

        if *SC == green*

                for $\forall \perp TL$

                        if $\perp TL.RD < SD \rightarrow \perp TL.RD = SD$ *// set the duration of the red signal*

                        if $\perp TL.RD > SD \rightarrow max = \mathbf{max}(\forall \parallel TL.GD)$

                                if $max > \perp TL.RD \rightarrow \perp TL.RD = max$ *// set the duration of the red signal*

                end

        if *SC == red*

                for $\forall \perp TL$

                        if $\perp TL.RD < SD \rightarrow \perp TL.GD = SD$ *// set the duration of the green signal*

                        if $\perp TL.GD < SD \rightarrow min = \mathbf{min}(\forall \parallel TL.RD)$

                                if $min > \perp TL.GD \rightarrow \perp TL.GD = min$ *// set the duration of the green signal*

                end

end

---

6 Research of Algorithms for Control of Traffic Lights

The study of the ATCS algorithm was implemented using the developed software model. This study included conducting experiments simulating the same situation on a given road section to test several traffic light algorithms. The previously mentioned STLCS and ATLCA have been selected as comparative algorithms for controlling traffic lights for IATCS. The experiment input is the set number of vehicles generated every 2 s (flow rate) with random routes. The simulation road sections are two symmetrical schematic intersections between which there is a limited distance section of the road. There is a three-lane road on each side of the intersections, where traffic lights of the «left arrow «and» round traffic light type (clause 6.1 of the RF Traffic Rules) regulate two co-directional lanes.

The outputs of the experiment are:

1. The time it took the model to resolve the traffic situation. This indicator shows the efficiency of traffic light algorithms, based on the value of the time that the model will spend until all vehicles stop moving and the road is empty.
2. The number of congestion at the intersection. This indicator records cases when vehicles cannot fit on a limited road section, and they are forced to stay at an intersection (Fig. 10). This circumstance on the roads is a violation of paragraph 13.2 of the Traffic Rules of the Russian Federation. However, in real traffic conditions, this situation often arises and leads to congestion.

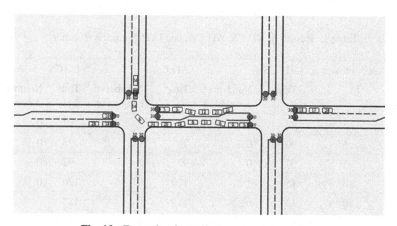

Fig. 10. Example of a traffic jam at an intersection

On the roads, a situation is possible when traffic stops altogether. This situation is typical if there are four vehicles from different directions, and because of a traffic jam at an intersection, each vehicle is not able to yield to another, acting according to the algorithm of the vehicle movement of the software model (Fig. 11). In this case, it makes no sense to measure time. If this situation arises, then it reflects the inefficiency of the traffic light regulation algorithm.

Twelve experiments were conducted. Table 1 displays the results of the study. If there was a situation leading to a complete cessation of movement (Fig. 11), then in the column "Time" put "–", the time when it happened was indicated in brackets.

In experiments 1-3 at low traffic intensity, each of the algorithms successfully coped with its task. This result is explained by the fact that traffic lights' algorithms were tested in conditions of not obstructed traffic. A minimal number of vehicles were recorded on the roads. With the STLCS algorithm, a sufficiently long interval of the green signal lighting duration allows vehicles to cross many intersections without obstruction. In this case, the identical duration of the red light is the only factor influencing road traffic. ATLCA and IATCS show the same data at low traffic volumes, i.e., in the absence of obstacles to

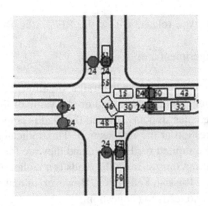

Fig. 11. An example of a complete cessation of movement.

Table 1. Results of STLCS, ATLCA, and IATCS models research.

Experiment number	Flow rate	STLCS		ATLCA		IATCS	
		Time (s)	Number of congestion	Time (s)	Number of congestion	Time (s)	Number of congestion
1	5	87	0	88	0	87	0
2	5	90	0	75	0	75	0
3	5	97	0	97	0	97	0
4	10	150	1	120	0	120	0
5	10	145	0	119	0	128	0
6	10	145	0	130	0	125	0
7	15	145	0	125	0	143	0
8	15	266	6	187	2	188	1
9	15	161	1	148	0	155	0
10	20	- (144)	1	- (133)	3	228	0
11	20	- (150)	1	188	2	213	0
12	20	- (112)	1	- (120)	1	178	0

vehicles' movement on a section of the road of limited length. ATLCA and IATCS adapt to the flow rate and change accordingly, the green light's duration before it turns on. Since the traffic intensity in these wholesalers is low, there are no congestions at intersections. Therefore, the algorithms work similarly in low traffic conditions. Experiments 2, 4–6, 8, 9 determined the algorithms' operation with an increased traffic intensity on the roads. These experiments demonstrate that ATLCA and IATCS performed better than STLCS. In 7 experiments, IATCS showed slightly worse results than ATLCA, but this is an individual case. Experiments 10, 12 determined the operation of the algorithms with a further increase in intensity. These experiments showed that STLCS stopped effectively solving the road situation, and each time the movement stopped (Fig. 11).

ATLCA also admitted a traffic interruption situation on the road. However, IATCS has shown the best results. The IATCS did not allow for a traffic jam at the intersection and therefore stop traffic. If ATLCA does not encounter the above situations, IATCS loses its advantage over it, as shown in experiment 1. This circumstance is explained by the fact that IATCS tries to avoid congestion, for which it increases the duration of the red signal burning from those directions from which it is worth allowing movement. In the ATLCA, there are no instructions for the occurrence of a traffic jam: vehicles, both allowing traffic jams will be able to continue driving, and everything will end successfully, or they will find themselves in a situation in which traffic is impossible (Fig. 11).

7   Conclusion

The study found that if the flow rate is low, then STLCS is an acceptable algorithm and can effectively cope with its task. As traffic increases, it is more expedient to switch to ATLCA. ATLCA unloads the traffic flow faster and more selectively, and traffic light signals do not light up "idle", as is the case in STLCS. However, when traffic is very high, the ATLCA can tolerate junction congestion if there are limited road segments. As a result, intractable traffic situations arise. However, the author's IATCS algorithm can cope with such a problem. Based on this, it makes sense to implement IATCS in the real process. The IATCS has proven to be effective in handling high flow rates at intersections with limited spacing.

References

1. Takashi, S., et al.: Construction of autonomous traffic light offset control system using multi agent system. Trans. Jpn. Soc. Artif. Intell. **26**(2), 324–329 (2011)
2. Biswas, S.P., Roy, P., Patra, N., Mukherjee, A., Dey, N.: Intelligent traffic monitoring system. In; Proceedings of the Second International Conference on Computer and Communication Technologies, vol. 380, pp. 535–545 (2015)
3. Ata, M.M., El-Darieby, M., Abd Elnaby, M., Napoleon, S.A.: Towards computer vision-based approach for an adaptive traffic control system. The Imaging Science Journal **66**(7), 419–432 (2018)
4. Zhang, W., Tan, G., Ding, N., Wang, G.: Traffic congestion evaluation and signal control optimization based on wireless sensor networks: model and algorithms. Mathematical Problems in Engineering 2012 (2012)
5. Chandel, S., Yadav, Y.: Modern traffic control system. Malaya J. Matematik **S**(1), 22–25 (2018)
6. Zavadsky, A.A., Shut, V.N.: Traffic management at intersections using multi-agent systems. Bull. Kherson National Tech. Univ. **3**(58), 90–98 (2016). (in Russian)
7. Agureev, I.E., Kretov, A.Yu., Matsur, I.Yu.: Investigation of algorithms for traffic light regulation of an intersection with different parameters of the traffic flow. Bull. Tula State University. Technical science **7–2**, 54-61 (2013). (in Russian)
8. Yousef, K.M.A., Al-Karaki, J., Shatnawi, A.M.: Intelligent traffic light flow control system using wireless sensors networks. J. Inf. Sci. Eng. **26**, 753–768 (2010)

9. Tubaishat, M., Shi, H., Yi, S.: Adaptive Traffic Light Control with Wireless Sensor Networks (2007). https://www.researchgate.net/profile/Malik_Tubaishat/publication/228613082_Adaptive_Traffic_Light_Control_with_Wireless_Sensor_Networks/links/0fcfd513a11966c636000000.pdf. Accessed 25 May 2019
10. Ghazal, B., Khatib, K., Khaled, C., Kherfan, M.: Smart traffic light control system. In: 2016 Third International Conference on Electrical, Electronics, Computer Engineering and their Applications (EECEA), pp. 161–166 (2016)
11. Kretov, A.Yu.: Review of some adaptive algorithms for traffic light regulation of intersections. Bulletin of the Tula State University. Techn. Sci. **7–2**, 61-67 (2013). (in Russian)
12. Zinoviev, I.V.: Algorithm for adaptive control of traffic lights based on the particle swarm algorithm. Sci. J. **4**(5), 26–30 (2016). (in Russian)

Author Index

Printed in the United States
By Bookmasters